AF508669

Renegotiating Disciplinary Fields in the Life Sciences

Renegotiating Disciplinary Fields in the Life Sciences

Editor

Alessandro Minelli

MDPI • Basel • Beijing • Wuhan • Barcelona • Belgrade • Manchester • Tokyo • Cluj • Tianjin

Editor
Alessandro Minelli
University of Padova
Italy

Editorial Office
MDPI
St. Alban-Anlage 66
4052 Basel, Switzerland

This is a reprint of articles from the Special Issue published online in the open access journal *Philosophies* (ISSN 2409-9287) (available at: https://www.mdpi.com/journal/philosophies/special_issues/renegotiating_disciplinary_fields_in_life_sci).

For citation purposes, cite each article independently as indicated on the article page online and as indicated below:

LastName, A.A.; LastName, B.B.; LastName, C.C. Article Title. *Journal Name* **Year**, *Volume Number*, Page Range.

ISBN 978-3-0365-0124-6 (Hbk)
ISBN 978-3-0365-0125-3 (PDF)

Cover image courtesy of Alessandro Minelli.

Contents

About the Editor

Alessandro Minelli (Full Professor of Zoology at the University of Padova until retirement in 2011) served as Vice-President of the European Society for Evolutionary Biology. For several years, his research focus was biological systematics, but since the mid-1990s, his research interests have turned towards evolutionary developmental biology and the philosophy of biology. He is the author of 'Biological Systematics' (1993), 'The Development of Animal Form' (2003), 'Forms of Becoming' (2009), 'Perspectives in Animal Phylogeny and Evolution' (2009) and 'Plant Evolutionary Developmental Biology' (2018).

Preface to "Renegotiating Disciplinary Fields in the Life Sciences"

Recent and ongoing debates in biology and the philosophy of biology reveal a widespread dissatisfaction with traditional explanatory frameworks. This is, for instance, the case of Neo-Darwinism, as it has been frequently advocated that evolutionary biology should replace the traditional gene-centered perspective with an organism-centered extended evolutionary synthesis, to account, e.g., for inclusive inheritance extending beyond genes and for phenotypic variation resulting from nonrandom mutation or biased by developmental processes. There are also problems with the current definitions or circumscriptions, often vague or controversial, of key concepts such as gene, species, and homology, and even of whole disciplinary fields within the life sciences, like developmental biology. To some extent, growing awareness of these conceptual issues and the contrasting views defended in their regard can be construed as marks of healthy debates in the field; however, this is also arguably a symptom of the need to revisit traditional, unchallenged partitions between the specialist disciplines within the life sciences. In the diversity of topics addressed and approaches to move beyond the current disciplinary organization, this Special Issue will hopefully stimulate further exploration towards an improved articulation of life sciences at the service of both science and philosophy.

Alessandro Minelli

Editor

 *philosophies*

Editorial

Renegotiating Disciplinary Fields in the Life Sciences

Alessandro Minelli

Department of Biology, University of Padova, Via Ugo Bassi 58B, I35131 Padova, Italy;
alessandro.minelli@unipd.it; Tel.: +39-389-949-4954

Received: 9 September 2020; Accepted: 3 December 2020; Published: 7 December 2020

The general problem around which this Special Issue revolves is that the way in which science is organized into specialties can have negative consequences on the progress of knowledge. Specifically, research priorities in biology and the circumscription of core concepts, both in biology and in the philosophy of biology, are very sensitive to the articulation of the Life Sciences in specialties.

The problems deriving from an inadequately critical attitude towards this issue affect biology as a whole due to the widespread lack of critical attitude, both in defending traditional disciplinary areas and in promoting new, or newly characterized and newly named, areas of research, often with an aggressive strategy.

In a well-documented and incisive article entitled 'Inclusion and Exclusion in the History of Developmental Biology', Nick Hopwood [1] demonstrates how the articulation of a science in specialties impinges on decisions on what the important problems are and how these must be addressed. This conditioning has a social dimension, as the division of a science into specialized disciplines very strongly affects the identity of a scientific community, and consequently the strategies of academic affirmation, and the criteria for the allocation of funds and the organization of undergraduate degree programs (see also [2,3]).

As soon as we become aware of the issue, action becomes possible. To use Hopwood's words [1] (p. 1), 'Disciplines are made, not found.' Quite a few disciplines, indeed, are simply defined on the basis of inclusion or exclusion criteria. This is more frequent in the case of ancient disciplines and those of an applied nature, e.g., in the domains of medicine and agriculture. There is nothing to blame, from an operational point of view, if individual researchers or institutions (including scientific societies and their journals) address sets of biological phenomena that only have in common the fact of dealing with the diseases of humans or domestic animals (human or veterinary pathology), or with crop plants, or aquatic animals relevant to fisheries. A very different thing, however, is to consider these disciplines as areas suitable for the development of general concepts, or theories, with regard to the living.

In the course of time, new disciplines emerge, generally characterized by a distinct set of problems or by a common technique, but often energetically pursuing less scientific targets such as the personal affirmation of a scholar or the creation of a new lobby aiming at success in the competition for funding and academic positions [1,4,5]. This must be seriously addressed if we wish to identify an organization of the biological disciplines that is able to stimulate and support the conceptual refreshment of the sciences.

To date, insufficient attention has been paid to the new perspectives that show up every time the boundaries between two or more disciplines are questioned or newly determined, often facilitating, in this way, the emergence of new questions, new research directions, and, in any case, helping to refresh notions and terms, including general and fundamental ones, which interest the philosopher no less than the biologist, such as individual, generation, development, reproduction, and evolution.

A collective effort is needed to move forward in this direction, which I hope the essays forming this Special Issue will contribute to. None of us are claiming to write a new biology, or a new philosophy of biology. Our pages, however, express an effort towards a vision of the living world that is both integrated and flexible in the identification of problems, concepts and their mutual relations.

Inevitably, it is difficult for each of us to avoid privileging the biological discipline to which we have devoted a lifetime of research. Even a scholar of the stature of Ernst Mayr, one of the most prestigious figures in evolutionary biology of the last century, insisted that there could not be a biology as a unitary science before the acceptance of an evolutionary vision of the living [6]—a questionable statement that neglects the importance of the cellular theory, as formulated by Schwann [7] ca. 20 years before the *Origin* [8]: 'it may be asserted, *that there is one universal principle of development for the elementary parts of organisms, however different, and that this principle is the formation of cells.* […] The development of the proposition, that there exists one general principle for the formation of all organic productions, and that this principle is the formation of cells, as well as the conclusions which may be drawn from this proposition, may be comprised under the term cell-theory.' ([9]; pp. 165–166; italics as in the original). The insistence on the role of evolutionary theory as a unifying principle of biology also overshadows another great merit of Charles Darwin, naturalist extraordinaire, who developed his works, including the *Origin*, on a documentary basis, without taxonomic restrictions.

This Species Issue includes five contributions. In their diversity, both the topics addressed and the approaches adopted to address them, these articles will hopefully stimulate further exploration towards an improved articulation of the Life Sciences at the service of both science and philosophy.

My main intended message in 'Disciplinary Fields in the Life Sciences: Evolving Divides and Anchor Concepts' is that advances in both biology and the philosophy of biology will benefit from a degree of flexibility in the way in which problems' agendas [10] are organized around (and through) concepts: this may require adopting unconventional perspectives and arguably, as suggested by other articles in this collection, a degree of pluralism too.

Indeed, Igor Pavlinov defends in his essay, a 'Multiplicity of Research Programs in the Biological Systematics: A Case for Scientific Pluralism'. Moving from a general philosophical position dominated by ideas of contemporary conceptualism and evolutionary epistemology, Pavlinov defends taxonomic pluralism. Biological diversity explored by biological systematics is a complex yet organized natural phenomenon that can be partitioned into several aspects, defined naturally with reference to various causal factors. These aspects are studied by research programs based on specific taxonomic theories. According to Pavlinov, each taxonomic theory is characterized by a unique combination of interrelated ontological and epistemological premises, which are most adequate to address one aspect of biological diversity. Phenetic, rational, numerical, typological, biosystematic, biomorphic, phylogenetic, and evo-devo research programs in systematics are recognized. From a scientific pluralism perspective, all of these research programs are of the same scientific significance, as possible parts of a generalized faceted classification.

The main message of Chris Fields and Michael Levin's paper titled 'How Do Living Systems Create Meaning?' is that the fundamental goal of Life Sciences is to understand how living systems create meaning. The authors discuss this question at multiple scales, from molecular interaction networks to the biosphere. Their focus is on identifying and understanding the roles of the reference frames that systems use, at each scale, to interpret their inputs. This involves understanding attention, memory, and the representation of *self* at multiple scales. From the perspective of disciplinary boundaries, Fields and Levin try to dissolve the one between biology and cognitive science.

In 'Evo-Devo: An Ongoing Revolution?', Salvatore Ivan Amato discusses the epistemological status of Evo-Devo as a theory. Evo-Devo has been considered as a new paradigm, a new research program or a revolutionary science within biology, but the variety of stances within this field makes it difficult to establish whether we are facing a new evolutionary synthesis. What is desirable is an adequate problematization of development, which can help clarify the relationship with both evolution and inheritance and take it one step further in the search for the identity of Evo-Devo.

Rolf Rutishauser revises 'The Past and Future of Continuum Plant Morphology' from a process-thinking perspective that allows to perceive and interpret growing plants as developmental continua, as process combinations rather than as assemblages of structural units ("organs") such as roots, stems, leaves, and flowers. This philosophical perspective was already favored by Agnes Arber (e.g., in her book on *The Natural Philosophy of Plant Form* [11]), and by Rolf Sattler, who first proposed his continuum approach in plant morphology exactly 50 years ago. This dynamic approach, known as continuum plant morphology, allows developmental geneticists and evolutionary biologists to move towards a more holistic understanding of multicellular plants in time and space.

Funding: This research received no external funding.

Acknowledgments: I express my gratitude to *Philosophies*, for offering me the opportunity to edit this Special Issue, and to the authors of the four articles that follow in this collection.

Conflicts of Interest: The author declares no conflict of interest.

References

1. Hopwood, N. Inclusion and exclusion in the history of developmental biology. *Development* **2019**, *146*, dev175448. [CrossRef] [PubMed]
2. Maienschein, J. Shifting assumptions in American biology: Embryology, 1890–1910. *J. Hist. Biol.* **1988**, *14*, 89–113. [CrossRef]
3. Sapp, J. The struggle for authority in the field of heredity, 1900–1932: New perspectives on the rise of genetics. *J. Hist. Biol.* **1983**, *16*, 311–342. [CrossRef] [PubMed]
4. Sapp, J. *Beyond the Gene. Cytoplasmic Inheritance and the Struggle for Authority in Genetics*; Oxford University Press: Oxford, UK, 1987; ISBN 0195042069.
5. Bowler, P.J. *The Non-Darwinian Revolution. Reinterpreting a Historical Myth*; John Hopkins University Press: Baltimore, MD, USA, 1988; ISBN 0801843677.
6. Mayr, E. *The Growth of Biological Thought. Diversity, Evolution and Inheritance*; Belknap Press of Harvard University Press: Cambridge, MA, USA, 1982; ISBN 0674364457.
7. Schwann, T. *Mikroskopische Untersuchungen über die Uebereinstimmung in der Struktur und dem Wachsthum der Thiere und Pflanzen*; Sander'sche Buchhandlung (G.E. Reimer): Berlin, Germany, 1839.
8. Darwin, C. *On the Origin of Species by Means of Natural Selection, or the Preservation of Favored Races in the Struggle for Life*; John Murray: London, UK, 1859.
9. Schwann, T. *Microscopical Researches into the Accordance in the Structure and Growth of Animals and Plants*; Smith, H., Translator; Seydenham Society: London, UK, 1847.
10. Love, A.C. Explaining evolutionary innovations and novelties: Criteria of explanatory adequacy and epistemological prerequisites. *Philos. Sci.* **2008**, *75*, 874–886. [CrossRef]
11. Arber, A. *The Natural Philosophy of Plant Form*; Cambridge University Press: Cambridge, UK, 1950.

 *philosophies*

Review

Disciplinary Fields in the Life Sciences: Evolving Divides and Anchor Concepts

Alessandro Minelli

Department of Biology, University of Padova, I 35131 Padova, Italy; alessandro.minelli@unipd.it;
Tel.: +39-389-949-4954

Received: 10 October 2020; Accepted: 3 November 2020; Published: 4 November 2020

Abstract: Recent and ongoing debates in biology and in the philosophy of biology reveal widespread dissatisfaction with the current definitions or circumscriptions, which are often vague or controversial, of key concepts such as the gene, individual, species, and homology, and even of whole disciplinary fields within the life sciences. To some extent, the long growing awareness of these conceptual issues and the contrasting views defended in their regard can be construed as a symptom of the need to revisit traditional unchallenged partitions between the specialist disciplines within the life sciences. I argue here that the current relationships between anchor disciplines (e.g., developmental biology, evolutionary biology, biology of reproduction) and nomadic concepts wandering between them is worth being explored from a reciprocal perspective, by selecting suitable anchor concepts around which disciplinary fields can flexibly move. Three examples are offered, focusing on generalized anchor concepts of generation (redefined in a way that suggests new perspectives on development and reproduction), organizational module (with a wide-ranging domain of application in comparative morphology, developmental biology, and evolutionary biology) and species as unit of representation of biological diversity (suggesting a taxonomic pluralism that must be managed with suitable adjustments of current nomenclature rules).

Keywords: nomadic concept; nomadic discipline; anchor concept; anchor discipline; life cycle; generation; organizational module; species

---

1. Introduction

The traditional articulation of biology into main research domains is not always satisfactory. For example, delimiting reproduction from other biological processes is not as easy as it might seem [1]. Let us focus on a strawberry plant. This produces runners that elongate horizontally on the ground, behaving like the branches of a growing plant, but after some time roots and leaves sprout on their lower face: a new plant takes shape, which ends up separating from the parent when the runner connecting them dries up. To use the terms introduced by John L. Harper and James White [2], even in the presence of a single *genet* (genetic individual), the production of runners leads to the formation of a new *ramet*, an anatomically separate individual, which therefore is the outcome of a reproductive event. Is this thus growth, or reproduction, or both?

Another difficult border separates developmental biology from some chapters of physiology, for example the physiology of nutrition or metabolism. In large snakes, e.g., pythons, in conditions of prolonged fasting the intestine undergoes a morphological and functional regression [3]; after a meal, cell proliferation reactivates the intestinal epithelium, which resumes its organization and functionality. This is accompanied by a rapid change in gene expression [4]. In terms of the mechanism, this is development; in terms of the function, this concerns the physiology of nutrition.

Cellular metabolites can modulate the activity of epigenetic factors that establish functional links between nutrition and gene expression [5,6]. More generally, intricate connections between anabolic processes and developmental transitions have been discovered [7,8].

As discussed below, recent and ongoing debates in biology and in the philosophy of biology reveal widespread dissatisfaction with the current definitions or circumscriptions, often vague or controversial, of key concepts such as gene, individual, species, and homology, and even of whole disciplinary fields within the life sciences. To some extent, the long growing awareness of these conceptual issues and the contrasting views defended in their regard can be construed as a symptom of the need to revisit traditional, unchallenged partitions between the specialist disciplines within the life sciences.

The problems deriving from an inadequately critical attitude towards the disciplinary partitions affect biology as a whole because of the unique diversity of the biological phenomena and the ubiquitous contrast between the inertial tendency to defend traditional disciplinary areas and a lively and often aggressive tendency to promote new or newly characterized and newly named research fields. The present paper is intended to contribute to a refreshment of this conceptual area. I articulate my argument as follows.

In the next Section, I focus on concepts as units of scientific knowledge that "continuously undergo transformation, and [...] function by *guiding ongoing scientific practice. A biology concept can motivate future* scientific efforts, and it can also provide a scaffold to direct the generation of new knowledge and the organization of complex knowledge." [9] (p. 82, italics as in the original). I offer examples of concepts that change meaning according to the disciplinary context in which they are discussed and disciplined, including intentionally created 'hybrids' [10].

In Section 3, I suggest one of the possible strategies we could adopt in revising the disciplinary structure of biology. The current relationships between the traditional disciplines and a number of core concepts that change meaning while used in widely different disciplinary contexts is worth being explored from a reciprocal perspective, by selecting suitable *anchor concepts* around which disciplinary fields can flexibly move. Three tentative examples are offered. The first example suggests *redefined concepts of generation* as units in a periodization of the life cycles that opens new perspectives on both development and reproduction, and their evolution. The second example focuses on a notion of *organizational module* flexible enough as to be equally of use in comparative morphology, developmental biology and evolutionary biology, without the requirement for morphological, developmental and evolutionary modules to be overlapping or hierarchically nested. The third example suggests adopting a similarly flexible notion of *species as unit of representation of biological diversity*, as anchor species within which the different species notions can be accommodated in a disciplined form of pluralism.

2. Questionable Boundaries between Biological Disciplines

2.1. Multidisciplinary or Interdisciplinary?

In biology, as in other sciences, interdisciplinarity has been steadily increasing in the last decades, but with a diversity of levels, intensity and outcome. The transfer of concepts, problems, and tools between disciplines is often strongly polarized, with the receiving discipline adopting them from a donating discipline [11], but often it is a two-way affair. In this case we can characterize the process as one of integration, either epistemic or organizational, or both [11]. Moreover, it is an accepted notion that in modern science many key concepts are shared by traditionally separate disciplines, and these are often concepts that do not lend themselves to precise definitions [12].

The biological concepts whose definition has proved more problematic and is still controversial are probably those of species, homology, gene, and individual. In the first case the controversy is particularly strong within the single biological discipline of systematics [13]; in the second case it involves different disciplines (morphology, phylogenetics etc.), mostly insofar as that these are united by the adoption of the comparative method [14]; see [15] for a broader perspective on these cases. The definition of gene has an overt transdisciplinary value, involving genetics in its various declinations, evolutionary biology, developmental biology, and the philosophy of biology [16–19]; the same applies for the definition of individual [20–25].

The need to address seriously, in a flexible and pluralistic way, the problem of a re-determination of the boundaries between biological disciplines is demonstrated by the number of concepts that in recent decades have assumed the value of nomadic concepts [26,27]. This term was proposed to describe concepts for which the meaning and domain of application changes with the new contexts into which they migrate. This has soon proved true also of the very notion of the nomadic concept [28–30]. I will use it to describe concepts that seems to be flexible enough to serve an epistemic role in different disciplinary contexts, but risk taking ever changing and not necessarily overlapping meanings.

It is legitimate to think that the lack of shareable definitions for the terms listed in the penultimate paragraph is not only a consequence of progress in the disciplines in which each of them originated, but also evidence of the disputable delimitation of biological disciplines. Eventually, we must acknowledge the historical specificity of individual disciplines, and possibly also the historical specificity of our own concepts of discipline [31] (p. 51).

2.2. Hybrid Disciplines—The Case of Evolutionary Developmental Biology

Gross conceptual rearrangements may be required by the emergence of a new hybrid discipline. This happened, in the last two decades of the 20th century, at the interface between evolutionary biology and developmental biology.

Interestingly, these two disciplines had been diverging more and more during most of the century. In the mind of authoritative evolutionary biologists, development was a black box between genotype and phenotype whose content could be ignored [32]. On the other hand, the transition from the descriptive embryology of the 19th century (which had provided valuable contributions to the understanding of phylogenetic relationships) to the experimental embryology of the following century had seen a progressive loss of interest in the comparative aspects of developmental processes and a growing focus on experimental work restricted to a very small number of model species [33,34]. Eventually, however, formidable technical advances in the second half of the 20th century made it possible to implement a developmental genetics program. One of the most sensational results was the discovery of the involvement of homologous genes in the development of such different organisms as mouse and fruit fly. The increasingly accessible contents of the black box between genotype and phenotype proved to be of utmost interest not only for development biologists, but also for evolutionary biologists. The emergence of a new research field in this interface area is conventionally fixed by two books whose publication dates and titles respectively mark the completion of the maturation phase and the first full expression of the new discipline. In 1983, Rudy Raff and Thomas Kaufman published a book [35], the title of which (*Embryos, Genes, and Evolution*) clearly identified the subject, approach, and problems of this discipline, while *Evolutionary Developmental Biology*, the title of the book published nine years later by Brian K. Hall [36], provided the name (often abbreviated as evo-devo) by which the latter was definitively identified [37,38].

Gilbert and Burian's early summary [39] that evo-devo "is both a synthesis between evolutionary biology and developmental biology and an ongoing negotiation between these two disciplines" (p. 61) is still up-to-date [40]. Winther [41] proposed to characterize evolutionary developmental biology as a *trading zone*, a catching term introduced by Galison [42] to indicate those interdisciplinary areas in which specific adoption and redefinition of both the concepts and the environment where they are implemented allow successful conceptual transfers [12,30].

For sure, conflicts cannot be avoided even in the best trading zones. In the case of evolutionary developmental biology, Love [43] has pointed to a hardly erasable tension between the two 'souls'. Repeatability of experimental results requires the highest possible uniformity in the strains (often long inbred laboratory lines) used in the tests. But this choice potentially deletes all the intraspecific variation on which evolution is deemed to happen. In this tug of war between developmental biology and evolutionary biology, it is not surprising that studies effectively coupling developmental genetics and population genetics (devgen-popgen [44]) are still few, despite a few excellent examples such as the already classic studies on the beaks of Darwin's finches revisited from an evo-devo perspective [45,46]).

2.3. Beyond Hierarchies and Facile Interdisciplinary Transfers

In these years that witness so much talk about the need to broaden the traditional neo-Darwinian vision of evolution to move towards an extended synthesis [47–49] it is necessary, in my opinion, to make an even more generous and adventurous effort and to seek, in an ever wider trading zone, to refresh the relationships between biological disciplines.

An overarching question is the relationship between the disciplinary articulation of the natural sciences and a hierarchical vision of nature structured into levels of organization.

The starting point of a debate that continues to this day [50,51], with an unceasing renewal of points of view and arguments, is a 1958 article by Oppenheim and Putnam [52] postulating both an articulation of nature in terms of levels of organization, and a close reciprocal correspondence between levels of organization in nature and the sciences that describe their components and develop theories on the relationships between them [53]. From these premises, Oppenheim and Putnam derived an entire reductionist program.

A description of reality in terms of levels of organization seemed useful even to those who were ready to recognize that theories are not always "limited to single levels, that levels are always well defined, or that two or more entities can always be unambiguously ordered with respect to level" [54] (p. 215). But in more recent times, the progressive move away from the reductionist program of Oppenheim and Putnam was one of the reasons for the decreasing favor of the very notion of levels of organization [55–57], until its total rejection by some authors [58]. Others have brought their arguments against these dismissionary positions, while nevertheless proposing new interpretations (and new epistemic roles) for the notion of organizational levels. While rejecting an ontological interpretation according to which the world would be structured by levels of organization, Brooks and Eronen [59]; see also [60,61] nevertheless save this notion as useful in the abstract description of systems and as a guide in the search for new areas of investigation to be explored. On the other hand, DiFrisco [62,63] rejects the criteria thus far used in identifying organization levels, in terms of compositional relationships or spatial scale, and suggests a dynamic approach that recognizes levels defined on the basis of rates or time scales of processes. Baedke [64] challenges the general acceptance of a never changing existence of levels of organization such as cells, tissues, organs, and individual organisms and points to the necessity of addressing their dynamical nature over developmental time and in evolution.

An overlooked consequence of the generalized acceptance of a vision of the living world in terms of compositional levels organized in part-whole relations [65] is the creation of disciplines through a copy-and-paste process. If in the study of humans and, more generally, of animals, it has proved useful to recognize a science of cells (cytology), a science of tissues (histology), a science of embryonic development (embryology) etc., this disciplinary articulation was accepted as sensible for all animals and even for multicellular organisms at large and corresponding disciplines were created for plants. Many biologists may take for granted, for example, the legitimacy of a plant embryology, but this should be resisted. In plant science, the use of the term *embryo* for the future seedling still enclosed within the seed casings was virtually unknown until 1788, when Gaertner [66] successfully introduced it in his treatise *De fructibus et seminibus plantarum* (On Plant Fruits and Seeds). Gaertner borrowed a number of terms and concepts from animal embryology. Some of these terms have remained in use for both plants and animals, but nobody would venture today to say, for example, that the placenta of plants is homologous to the placenta of mammals. Unfortunately, instead, the idea of an equivalence between what is called an embryo in either kingdom is still widespread, even among professionals [67].

3. Moving Ahead—Nomadic Concepts or Nomadic Disciplines?

I mentioned above that several core concepts of the life sciences have been taking continuously new meanings as long as their domain of application has been shifting from one biological discipline to another. Other adjustments to the meaning of concepts have accompanied the emergence of new subdisciplines within an older research area in which the concept was already in use, but with a different meaning. In a dialectic relationship between concepts and disciplines, the former have been continuing their *nomadic* existence, taking different meanings as a consequence of their changing association with disciplines, each of which acts as a semantic *anchor context*.

Traditionally, disciplines are taken for granted, as *anchor disciplines*, and concepts may nomadically wander from one to another. My suggestion here is, that a reversed relationship between disciplines and core concepts may prove useful, at least as an epistemic tool to be used to refresh the traditional divides separating a number of biological disciplines. I am suggesting indeed that we should perhaps move from a few *anchor concepts* around which *nomadic disciplines* may continuously evolve.

In this reversed perspective, a suitably chosen concept becomes the anchor around which important problems agendas in a number of traditional disciplines can nomadically move. To be sure, within the general framework I am suggesting, one of disciplinary flexibility, no anchor concept shall be regarded as definitely fixed, but it may be worth exploring for a while its possible epistemic usefulness. I will offer here three examples.

3.1. Anchor Concept 1—Nomadic Disciplines in the Study of the Life Cycle

Much of biology deals with objects, processes, and concepts about reproduction, development, or evolution. A great many problems have been successfully addressed by restricting attention to one or another of these main themes and often accepting—operationally at least—variously restrictive and partial notions of reproduction, development, or evolution. These partitions, however, limit the formulation of potentially interesting questions and thus constrain progress both in the life sciences and in the philosophy of biology. Therefore, it may be sensible to explore in a free, innovative way the notoriously difficult and therefore challenging border areas between development and reproduction, and those between development and evolution. As mentioned above, the latter frontier came into the spotlight at the end of the last century and has been generally managed by making room for a new discipline, evolutionary developmental biology, within which new key concepts have emerged such as evolvability [68,69], modularity [70] and innovation [15,71–73]. However, a small number of scholars prefer a different path, suggesting a much closer integration between the two traditional disciplines [74,75].

The other boundary, the one between reproduction and development, may deserve a conceptual re-organization by treating these two chapters of biology as *nomadic disciplines* whose core problems vary according to their various association with a small number of *anchor core concepts*.

To the best of my knowledge, this reversal of perspective has not been formalized before; however, to see how this may operate, we can get advice from the literature. Let us start by comparing the complementary approaches of Paul E. Griffiths, Karola Stotz, and James Griesemer to the relationships between reproduction and development. On the one hand, Griffiths and Stotz take development as the core process, and reproduction as the linking one: "It is the developmental process that replicates itself across the generations" [76] (p. 227). A complementary proposal, centered instead on reproduction, comes from Griesemer, who takes reproducers as the units of multiplication, hereditary variation, and development: by reproduction, a parent generates an offspring that is able to develop so as to generate its own offspring [77]. Both perspectives have their merits and can suggest new, original research programs. In this context, it is fitting to remark that "Dobzhansky's declaration (1973) [78] that nothing in biology makes sense except in the light of evolution notwithstanding, it is equally clear that evolution does not make sense except in the light of the rest of biology [79]. So, it is a toss-up which concepts must "come first" in order to understand "the rest"" [80] (p. 140).

By introducing unprecedented perspectives on reproduction and development, these proposals contribute to the debates on the notion of the individual [22–27], the nature of development [81–83], and the diversity of hereditary mechanisms, which include Mendelian ones but are not limited to these [84–90]. These theoretical approaches put center stage the *biological cycle*, rather than the individual organism, both in ontogenetic and evolutionary perspectives [91–93].

Largely overlooked, however, is the problem of a segmentation (periodization) of biological cycles suitable for comparisons across the different branches of the tree of life. Current textbook descriptions do not help much [1]. For example, it is taken for granted that the biological cycle of most animals is monogenerational. Two generations are recognized only in particular instances, for example in those cnidarians in which an asexually reproducing polyp alternates with a sexually reproducing medusa. In plants, the haploid cells that derive from meiosis do not behave like gametes. Instead, each of them (a spore) gives rise to a multicellular organism. Two generations are also recognized in plants, a diploid sporophyte alternating with a haploid gametophyte. In the flowering plants, the sporophyte is the conventional individual plant, while the gametophyte is much less conspicuous: in the male version, the gametophyte is the pollen grain, made up of only three cells; in the female version, it is the complex of an egg plus an embryo sac of six cells.

To identify comparable generational units in the life cycles of organisms belonging to the different evolutionary lines, a suitable periodization is required. Gorelick [94] proposed to recognize the beginning of a new generation every time a sexual phenomenon brings about changes in the chromosomal set. In the typical biological cycle of animals, we should therefore recognize two generations, respectively initiated by meiosis (which leads from diploid to haploid condition) and karyogamy (the fusion of the haploid nuclei of the gametes, which reinstates diploidy). If we accept Gorelick's proposal, eggs and spermatozoa represent a haploid generation distinct from the diploid generation that begins with the zygote.

Gorelick's suggestion is very reasonable, irrespectively of whether or not it is advisable to retain the term generation for each of the two segments of an animal's life cycle. The real problem with this proposal is instead that it does not work for many groups, e.g., plants or ciliate protozoans. The articulation of the biological cycle into a haploid and a diploid generation separated from each other by meiosis and karyogamy works only for diplobiont organisms, i.e., those in which the cells that derive from meiosis are the gametes, and for haplobionts, i.e., those in which the diploid condition is restricted to the zygote, that directly undergoes meiosis, starting a new haploid generation, as in *Volvox* and other green algae [1]. But plants, as said, are haplodiplobiont. To dissect their life cycle into meaningful phases, Gorelik's periodization is insufficient.

To describe the different relationships between sexuality and reproduction in the biological cycle, irrespective of its haplobiont, diplobiont, or haplodiplobiont nature, I proposed [95] to recognize two kinds of generations (and, correspondingly, two kinds of individuals):

Demographic generation: the individuals produced by sexual or asexual reproduction by individuals of a parent generation.

Genetic generation: a set B of individuals produced by a set A of individuals (which represents a distinct genetic generation) by a sexual process (sexual reproduction or pure sexuality, i.e., sexuality without reproduction, as in the ciliate protozoans, see below).

Based on these definitions (but also based on Gorelik's proposal), in the life cycle of animals like humans there are two generations. More precisely, the haploid generation (gametes) is a demographic generation, while the diploid generation (the conventional individual organisms) is a genetic generation (beginning with karyogamy), but it is not a new demographic generation (fertilization does not increase population size).

In plants, however, the biological cycle includes three generations, separated either by a reproductive event (production of gametes without meiosis), by a sexual event (karyogamy followed by sporophyte development) or by overlapping sexual and reproductive events (production of spores through meiosis).

The case for ciliate protozoans is very different. In these single-cell organisms, there is no association between sexuality and reproduction. Sexuality consists here of the exchange of nuclei between two cells that retain their somatic identity and resume independent life after the exchange: two ex-conjugants are genetically 'renewed', but this has no demographic consequence; reproduction is achieved by simply dividing the diploid cell into two daughter cells, without any genetic change.

Therefore, in ciliates, genetic generations separated by meiosis and fertilization do not coincide with the demographic generations punctuated by mitoses.

A periodization of the biological cycle based on the notions of generation defined above opens the way to comparisons between the most diverse groups of living beings and suggests new perspectives, both for developmental biology and reproductive biology. For example, recognizing the meaning of generation (demographic and genetic) in the unicellular phase of a typical animal life cycle legitimizes the description of gametogenesis in terms of developmental biology (and therefore provides an extension of developmental biology to unicellular organisms). At the same time, the distinction between genetic generation and demographic generation leads to recognizing, in the biological cycle of animals, the exclusively genetic nature of the diploid generation, except in the case of polyembryony, that is to say, the production, through a reproductive event not accompanied by sexual phenomena, of two or more embryos (twins) starting from the same zygote [1].

3.2. Anchor Concept 2—Organizational Module

Most of research in biology presupposes a sensible decomposition of the complex body architecture of the individual into recognizable parts based on morphological or functional criteria. For example, a textbook tradition made solid by the general acceptance of a hierarchical vision of nature suggests an articulation of the body into organs, each of which is made up of variously differentiated tissues, formed in turn by cells. This hierarchy continues with the subcellular structures, which for simplicity, we can ignore here. For the same reason we will ignore the complications deriving from the interactions between the cells of an animal and those of the prokaryotes forming the associated microbiome [96–100]. Instead, we will address the question, whether it is possible to identify, in the structure of a multicellular organism, fundamental units on which we can focus in the most diverse disciplinary contexts: for example, as structural units in comparative anatomy (homologues), and as largely autonomous elements from the point of view of development or evolution.

A good starting point is Günter Wagner's revisitation, first formulated towards the end of the last century, of the concept of homology: "Structures from two individuals or from the same individual are homologous if they share a set of developmental constraints, caused by locally acting self-regulatory mechanisms of organ differentiation. These structures are thus developmentally individualized parts of the phenotype." [101] (p. 62). Subsequently, shifting the focus from development to evolution, Wagner stated that "homologues can be understood as modular units of evolutionary transformation" [102] (p. 36).

In the same years the use of 'module' as a neutral term (in a disciplinary sense) began to spread, to the point that Borja Esteve–Altava could affirm that "The study of modularity is commonplace in all biological disciplines" [103]. However, difficulties in reaching agreement on a definition of module did not take long to emerge [70,104,105].

A basic issue is the distinction between variational modules [106–110] vs. organizational modules [103]. A *variational module* is a set of traits that vary in a coordinated manner, in some way independent of other groups of traits within a given system, for example, within the same individual organism. These patterns of (co)variation can be useful for recognizing units worth focusing on, but this module concept is arguably too vague to serve as a useful anchor concept. More promising are the *organizational modules*. These can be defined as groups of elements that establish more and/or stronger interactions within the group than outside it. In this deliberately abstract definition, the nature of the interactions that define a module is not specified. The concept is therefore applicable, within biology, to the units on which the most diverse disciplines focus, facilitating comparisons and exchanges between them [103,106]. Organizational modules are, for example, the parts of the body recognized by the morphologist, based for example on relationships of contiguity and connection, or on the gene regulatory networks of developmental genetics, or on the developmental modules on which Wagner's biological notion of homology is based, or on the morphofunctional units (e.g., the circulatory system, heart, and brain) which the physiologist deals with, or on sets of traits that evolve in a coordinated way, whatever the cause, such as selection or concerted evolution [101–113].

Units defined from different perspectives do not necessarily overlap [114]. The correspondence between developmental modules and evolutionary modules is a big problem for which we do not have any general solution [8]. Characteristically, organs such as heart or brain are obvious morphological and functional units, but are not modules from a morphogenetic point of view: in other terms, there are, for example, hearts as modular organs, but not a 'cardiogenesis' as a correspondingly integrated and largely autonomous developmental process [115–117].

3.3. Anchor Concept 3   Species as Unit of Representation of Biological Diversity

Few of the most important concepts in biology have taken on the character of nomadic concepts for as long as the concept of species, to the point that the topic already turned into a 'species problem' more than a hundred years ago. When this expression appeared for the first time in the title of a book [118], the species problem had already been the subject of numerous articles, some of which date back to the early years of the last century [119–121]. An article in *Nature* that highlighted the importance of that book reported an impressive example of the consequences of the plurality of alternative species concepts supported by taxonomists at the time: "Some years ago […] I endeavored to get from my colleagues at the Museum [=The British Museum (Natural History), as it was called at the time] estimates of the numbers of species in the various groups with which they were specially conversant. Some of the answers obtained were very interesting. With regard to mammals I was told "anything from 3000 to 20,000, according to the view you take as to what constitutes a species"" [122] (p. 440).

Ninety years later, we are still witnessing an endless evolution of the species concept, as tagged in the title of a recent article [123]. More or less formally, no less than 30 different notions of species have been proposed to date [13,124].

Ironically, perhaps, it is precisely the substantial intractability of the 'species problem' the strongest stimulus to adopt a 'radical solution' that may turn the species into a veritable anchor concept. The way forward was suggested by Robert O'Hara towards the end of the last century: "Perhaps the species problem is not something that needs to be solved, but rather something that [...] needs to be gotten over" [125] (p. 232). In addressing the 'species problem', we do not face a problem of fact, but rather as a problem of systematic representation [125]. Indeed, in agreement with Robert O'Hara's revisitation (and eventual dismissal) of the 'species problem,' I suggest here that, rather than accepting the species concept as incurably nomadic, we can fix it by divesting it of all the special qualities that differentiate its all too many notions, to fix it simply as a *unit of systematic representation* of biological diversity. This simplified concept returns to the different disciplinary domains in which the term 'species' is used the burden to customize it as required, whereas taxonomy continues to accommodate all the taxonomic units that biology requires without introjecting the controversies about species concepts, their legitimacy, and the priorities, conflicts, and compatibilities among them.

In principle, within one and the same set of organisms we can recognize groups on the basis of the most diverse criteria, which may correspond to different concepts of species, e.g., biological species, on the basis of reproductive isolation; morphological species, based on structural similarities, and so on. In different contexts, one choice may be preferable to another, while the claim to recognize a criterion of absolute validity appears unsustainable. Among others, Mishler and Donoghue [126] consequently defended a certain degree of pluralism in biological classification. Among the positions expressed in the last three decades, some authors (for example, [127–130]) support pluralism, others (for example, [131,132]) reject it. Despite the intentions and efforts of many taxonomists, current taxonomic practice is already remarkably pluralistic [133]. This choice, however, can be sustained only if the various species concepts adopted in the different instances are stated explicitly or are at least evident from the context [134].

Moreover, this pluralism may require adjustments in nomenclature [127]. Changes in a group's taxonomy are in fact a source of ambiguity in the meaning of species names. For example, until a few years ago, almost all authors classified African elephants as belonging to a single species (*Loxodonta africana*); at most, some zoologists distinguished the forest elephant as a subspecies (*Loxodonta africana cyclotis*) distinct from the savannah elephant (*Loxodonta africana africana*). However, recent studies [135] have led to classifying the two forms as two distinct species, in which case the savannah elephant retains the name *Loxodonta africana*, while the forest elephant takes the name *Loxodonta cyclotis*.

Diverging (or renewed) opinions about the taxonomic treatment of a group (for example, how many species, and which ones, are best recognized within a genus) are current in zoology as well as in botany. This circumstance, combined with the (otherwise sensible) rules of the international codes of nomenclature [136,137] causes the names attributed to the species to become semantically unstable. To avoid misunderstandings, it is sometimes necessary to specify the *taxonomic concept* [138], that is, the precise meaning of the name as used in a particular work. For example, in the case of African elephants, *Loxodonta africana* as used by [135], where the name is applied to the savannah elephant only, is different from *Loxodonta africana* as used by [139], which refers to all African elephants, treated as a single species: the latter taxonomic concept includes the former, but it is not identical to it [127,133]. The number of different taxonomic concepts that can correspond to a single Linnaean name is sometimes impressive. Precise data are available regarding birds, where for approximately 10,000 species and 22,000 subspecies that are currently recognized, over 1.5 million taxonomic concepts are available in the literature [140,141].

4. Summary and Conclusions

Until well into the second half of the 20th century, biology was accorded a status dramatically lower than other sciences, physics in particular. Similarly, the philosophy of science of the time largely disregarded the life sciences, to concentrate instead on those more mathematized and more rich in theory. Things changed dramatically in recent decades. In the meantime, however, advances in scientific knowledge have fragmented biology into an ever-growing plurality of disciplines, each of which, at least for a time, has consolidated around more promising or more successful research agendas. Only in exceptional cases, such as at the interface between developmental biology and evolutionary biology, has a new discipline (evolutionary developmental biology) taken shape in which a debate around the founding concepts is still in progress.

Today, the life sciences are extensively populated by nomadic concepts that take on the most diverse meanings depending on the contexts in which they are recognized and used. This is possibly attractive for the philosopher of science, but problematic and potentially dangerous for the scientist.

It is legitimate to ask whether in our day, it is still useful and appropriate to address the conceptual problems of biology with the breadth of horizons of what two centuries ago took shape under the name of biology. Today, in fact, this name often identifies only a large container, in front of which it is not regarded necessary or useful to address critically and flexibly the possible relations between its contents. This situation is arguably acceptable to many, or most, practicing biologists. However, it is not so for the philosophy of biology, which up to now has taken into consideration only a part of the concepts and problems faced by the life sciences.

Eventually, both biology and philosophy need a refreshment of the reciprocal relations between the different disciplines recognizable in this field.

In the previous pages, I have suggested one of the possible strategies we could adopt in revising the disciplinary structure of biology. The current relationships between anchor disciplines such as comparative morphology, systematics, evolutionary biology, developmental biology, biology of reproduction, and nomadic concepts wandering between them is worth being explored from a reciprocal perspective, etc., by selecting suitable anchor concepts around which disciplinary fields can flexibly move.

Advantages and shortcomings of a re-organization of biology as a set of nomadic disciplines revolving around a small number of anchor concepts is a challenge whose results deserve careful evaluation. As suggested by a referee, another nomadic concept that may deserve being re-refined in such a way as to fix it as anchor concept is the concept of environment. This being already at the boundary between biology and many other sciences, and largely beyond my own field on enquiry, I must leave it to others to explore the heuristic potential of this attractive suggestion.

Funding: This research received no external funding.

Acknowledgments: I am very grateful to Jan Baedke, James DiFrisco, Giuseppe Fusco, James Griesemer, Alan C. Love, Rolf Rutishauser, and three anonymous referees for their constructive criticisms on a previous version of this article.

Conflicts of Interest: The author declares no conflict of interest.

References

1. Fusco, G.; Minelli, A. *The Biology of Reproduction*; Cambridge University Press: Cambridge, UK, 2019. [CrossRef]
2. Harper, J.L.; White, J. The demography of plants. *Ann. Rev. Ecol. Syst.* **1974**, *5*, 419–463. [CrossRef]
3. Andersen, J.B.; Rourke, B.C.; Caiozzo, V.J.; Bennett, A.F.; Hiàcks, J.W. Postprandial cardiac hypertrophy in pythons. *Nature* **2005**, *434*, 37–38. [CrossRef] [PubMed]
4. Andrew, A.L.; Card, D.C.; Ruggiero, R.P.; Schield, D.R.; Adams, R.H.; Pollock, D.D.; Secor, S.M.; Todd, A.; Castoe, T.A. Rapid changes in gene expression direct rapid shifts in intestinal form and function in the Burmese python after feeding. *Physiol. Genom.* **2015**, *47*, 147–157. [CrossRef]
5. Sassone-Corsi, P. When metabolism and epigenetics converge. *Science* **2013**, *339*, 148–150. [CrossRef] [PubMed]
6. Von Dassow, G.; Munro, E. Modularity in animal development and evolution: Elements of a conceptual framework for EvoDevo. *J. Exp. Zool. B Mol. Dev. Evol.* **1999**, *285*, 307–325. [CrossRef]
7. Nagaraj, R.; Sharpley, M.S.; Chi, F.; Braas, D.; Zhou, Y.; Kim, R.; Clark, A.T.; Banerjee, U. Nuclear localization of mitochondrial TCA cycle enzymes as a critical step in mammalian zygotic genome activation. *Cell* **2017**, *168*, 210–223. [CrossRef] [PubMed]
8. Song, Y.; Shvartsman, S.Y. Chemical embryology redux: Metabolic control of development. *Trends Genet.* **2020**, *36*, 577–586. [CrossRef]
9. Brigandt, I. How are biology concepts used and transformed? In *Philosophy of Science for Biologists*; Kampourakis, K., Uller, T., Eds.; Cambridge University Press: Cambridge, UK, 2020; pp. 79–101.
10. Østreng, W. Crossing scientific boundaries by way of disciplines. In *Complexity. Interdisciplinary Communications 2006/2007*; Østreng, W., Ed.; Centre for Advanced Study: Oslo, Norway, 2008; pp. 11–13.
11. Gerson, E.M. Integration of specialties: An institutional and organizational view. *Stud. Hist. Philos. Sci.* **2013**, *44*, 515–524. [CrossRef] [PubMed]
12. Müller, E. Interdisciplinary concepts and their political significance. *Contrib. Hist. Concepts* **2011**, *6*, 42–52. [CrossRef]
13. Zachos, F.E. *Species Concepts in Biology. Historical Development, Theoretical Foundations and Practical Relevance*; Springer: Basel, Switzerland, 2016; ISBN 978-3-3194-4966-1.
14. Minelli, A.; Fusco, G. Homology. In *The Philosophy of Biology: A Companion for Educators*; Kampourakis, K., Ed.; Springer: Dordrecht, The Netherlands, 2013; pp. 289–322. [CrossRef]
15. Wagner, G.P. *Homology, Genes, and Evolutionary Innovation*; Princeton University Press: Princeton, NJ, USA; Oxford, UK, 2014; ISBN 9780691156460.
16. Portin, P.; Wilkins, A. The evolving definition of the term "gene". *Genetics* **2017**, *205*, 1353–1364. [CrossRef]
17. Snyder, M.; Gerstein, M. Defining genes in the genomics era. *Science* **2003**, *300*, 258–260. [CrossRef]
18. Griffiths, P.E.; Stotz, K. Genes in the postgenomic era. *Theor. Med. Bioeth.* **2006**, *27*, 499–521. [CrossRef] [PubMed]
19. Müller-Wille, S.; Rheinberger, H.-J. *Das Gen im Zeitalter der Postgenomik. Eine Wissenschaftshistorische Bestandsaufnahme*; Suhrkamp: Frankfurt am Main, Germany, 2009; ISBN 9783518260258.
20. Santelices, B. How many kinds of individual are there? *Trends Ecol. Evol.* **1999**, *14*, 152–155. [CrossRef]
21. Wilson, J. *Biological Individuality: The Identity and Persistence of Living Entities*; Cambridge University Press: Cambridge, UK, 1999; ISBN 0521624258.
22. Godfrey-Smith, P. *Darwinian Populations and Natural Selection*; Oxford University Press: New York, NY, USA, 2009; ISBN 9780199552047.
23. Bouchard, F.; Huneman, P. (Eds.) *From Groups to Individuals. Evolution and Emerging Individuality*; MIT Press: Cambridge, MA, USA, 2013; ISBN 9780262018722.

24. Pradeu, T. Organisms or biological individuals? Combining physiological and evolutionary individuality. *Biol. Philos.* **2016**, *31*, 797–817. [CrossRef]

25. Fields, C.; Levin, M. Are planaria individuals? What regenerative biology is telling us about the nature of multicellularity. *Evol. Biol.* **2018**, *45*, 237–247. [CrossRef]

26. Stengers, I. (Ed.) *D'une Science a L'autre: Des Concepts Nomads*; Seuil: Paris, France, 1987; ISBN 8877570180.

27. Surman, J.; Stráner, K.; Haslinger, P. Nomadic concepts—Biological concepts and their careers beyond biology. *Contr. Hist. Concepts* **2014**, *9*, 1–17. [CrossRef]

28. Bal, M. *Travelling Concepts in the Humanities: A Rough Guide*; University of Toronto Press: Toronto, ON, Canada, 2002; ISBN 0802035299.

29. Wolfe, C.T. The organism as ontological go-between Hybridity, boundaries and degrees of reality in its conceptual history. *Stud. Hist. Philos. Biol. Biomed. Sci.* **2014**, *48*, 151–161. [CrossRef]

30. Surman, J.; Stráner, K.; Haslinger, P. Nomadic concepts in the history of biology. *Stud. Hist. Philos. Biol. Biomed. Sci.* **2014**, *48*, 127–129. [CrossRef]

31. Suárez-Diaz, E. Molecular evolution: Concepts and the origin of disciplines. *Stud. Hist. Philos. Biol. Biomed. Sci.* **2009**, *40*, 43–53. [CrossRef]

32. Laubichler, M.D. Evolutionary developmental biology offers a significant challenge to the neo-Darwinian paradigm. In *Contemporary Debates in the Philosophy of Biology*; Ayala, F., Arp, R., Eds.; Wiley-Blackwell: Malden, MS, USA, 2010; pp. 199–212. [CrossRef]

33. Jenner, R.A. Unburdening evo-devo: Ancestral attractions, model organisms, and basal baloney. *Dev. Genes Evol.* **2006**, *216*, 385–394. [CrossRef]

34. Minelli, A.; Baedke, J. Model organisms in evo-devo: Promises and pitfalls of the comparative approach. *Hist. Philos. Life Sci.* **2014**, *36*, 42–59. [CrossRef]

35. Raff, R.A.; Kaufman, T.C. *Embryos, Genes, and Evolution*; Macmillan: New York, NY, USA, 1983; ISBN 0253206421.

36. Hall, B.K. *Evolutionary Developmental Biology*; Chapman & Hall: London, UK, 1992. [CrossRef]

37. Love, A.; Raff, R.A. Knowing your ancestors: Themes in the history of evo-devo. *Evol. Dev.* **2003**, *5*, 327–330. [CrossRef]

38. Horder, T.J. A history of evo-devo in Britain. *Ann. Hist. Philos. Biol.* **2008**, *13*, 101–174.

39. Gilbert, S.F.; Burian, R.M. Development, evolution, and evolutionary developmental biology. In *Keywords and Concepts in Evolutionary Developmental Biology*; Hall, B.K., Olson, W., Eds.; Harvard University Press: Cambridge, MA, USA, 2003; pp. 61–68.

40. Baedke, J.; Gilbert, S.F. Evolution and development. In *The Stanford Encyclopedia of Philosophy*, Fall 2020 ed.; Zalta, E.N., Ed.; Metaphysics Research Lab., Stanford University: Sanford, CA, USA, 2020. Available online: https://plato.stanford.edu/archives/fall2020/entries/evolution-development/ (accessed on 7 August 2020).

41. Winther, R.G. Evo-devo as a trading zone. In *Conceptual Change in Biology: Scientific and Philosophical Perspectives on Evolution and Development*; Love, A.C., Ed.; Springer: Dordrecht, The Netherlands, 2015; pp. 459–482. [CrossRef]

42. Galison, P. *Image and Logic: A Material Culture of Microphysics*; University of Chicago Press: Chicago, IL, USA, 1997. [CrossRef]

43. Love, A.C. Idealization in evolutionary developmental investigation: A tension between phenotypic plasticity and normal stages. *Philos. Trans. R. Soc. Lond. Biol. Sci.* **2010**, *365*, 679–690. [CrossRef]

44. Abzhanov, A.; Protas, M.; Grant, B.R.; Grant, P.R.; Tabin, C.J. *Bmp4* and morphological variation of beaks in Darwin's finches. *Science* **2004**, *305*, 1462–1465. [CrossRef] [PubMed]

45. Abzhanov, A.; Kuo, W.P.; Hartmann, C.; Grant, B.R.; Grant, P.R.; Tabin, C.J. The Calmodulin Pathway and evolution of elongated beak morphology in Darwin's finches. *Nature* **2006**, *442*, 563–567. [CrossRef]

46. Gilbert, S. Evo-devo, devo-evo and devgen-popgen. *Biol. Philos.* **2003**, *18*, 347–352. [CrossRef]

47. Pigliucci, M.; Müller, G.B. (Eds.) *Evolution: The Extended Synthesis*; MIT Press: Cambridge, MA, USA, 2010. [CrossRef]

48. Laland, K.N.; Uller, T.; Feldman, M.; Sterelny, K.; Müller, G.B.; Moczek, A.; Jabonka, E.; Odling-Smee, J. Does evolutionary theory need a rethink? Yes, urgently. *Nature* **2014**, *514*, 161–164. [CrossRef]

49. Laland, K.N.; Uller, T.; Feldman, M.; Sterelny, K.; Müller, G.B.; Moczek, A.; Jabonka, E.; Odling-Smee, J. The extended evolutionary synthesis: Its structure, assumptions and predictions. *Proc. R. Soc. Lond. B* **2015**, *282*, 20151019. [CrossRef] [PubMed]

50. Eronen, M.I.; Brooks, D.S. Levels of organization in biology. In *The Stanford Encyclopedia of Philosophy*, Spring 2018 ed.; Zalta, E.N., Ed. Available online: https://plato.stanford.edu/archives/spr2018/entries/levels-org-biology/ (accessed on 9 August 2020).

51. Brooks, D.S.; DiFrisco, J.; Wimsatt, W.C. (Eds.) Introduction. In *Levels of Organization in the Biological Sciences*; MIT Press: Cambridge, MA, USA, forthcoming.

52. Oppenheim, P.; Putnam, H. Unity of science as a working hypothesis. In *Minnesota Studies in the Philosophy of Science*; Feigl, H., Maxwell, G., Scriven, M., Eds.; University of Minnesota Press: Minneapolis, MN, USA, 1958; pp. 3–36.

53. Brigandt, I. Beyond reduction and pluralism: Toward an epistemology of explanatory integration in biology. *Erkenntnis* **2010**, *73*, 295–311. [CrossRef]

54. Wimsatt, W.C. Reductionism, levels of organization, and the mind-body problem. In *Consciousness and the Brain. A Scientific and Philosophical Enquiry*; Globus, G.G., Maxwell, G., Savodnik, I., Eds.; Plenum: New York, NY, USA, 1976; pp. 205–267. [CrossRef]

55. Eronen, M.I. Levels of organization: A deflationary account. *Biol. Philos.* **2015**, *30*, 39–58. [CrossRef]

56. Eronen, M.I. No levels, no problems: Downward causation in neuroscience. *Philos. Sci.* **2013**, *80*, 1042–1052. [CrossRef]

57. Potochnik, A.; McGill, B. The limitations of hierarchical organization. *Philos. Sci.* **2012**, *79*, 120–140. [CrossRef]

58. Thalos, M. *Without Hierarchy: The Scale Freedom of the Universe*; Oxford University Press: Oxford, UK, 2013. [CrossRef]

59. Brooks, D.S.; Eronen, M.I. The significance of levels of organization for scientific research: A heuristic approach. *Stud. Hist. Philos. Biol. Biomed. Sci.* **2018**, *68–69*, 34–41. [CrossRef] [PubMed]

60. Brooks, D.S. A new look at 'levels of organization' in biology. *Erkenntnis* **2019**. [CrossRef]

61. Brooks, D.S. In defense of levels: Layer cakes and guilt by association. *Biol. Theory* **2017**, *12*, 142–156. [CrossRef]

62. DiFrisco, J. Time scales and levels of organization. *Erkenntnis* **2017**, *82*, 795–818. [CrossRef]

63. DiFrisco, J. Integrating composition and process in levels of developmental evolution. In *Levels of Organization in the Biological Sciences*; Brooks, D.S., DiFrisco, J., Wimsatt, W.C., Eds.; MIT Press: Cambridge, MA, USA, forthcoming.

64. Baedke, J. The origin of new levels of organization. In *Levels of Organization in the Biological Sciences*; Brooks, D., DiFrisco, J., Wimsatt, W., Eds.; MIT Press: Cambridge, MA, USA, forthcoming.

65. Wimsatt, W.C. *Re-Engineering Philosophy for Limited Beings: Piecewise Approximations to Reality*; Harvard University Press: Cambridge, MA, USA, 2007; ISBN 9780674015456.

66. Gaertner, J. *De Fructibus et Seminibus Plantarum*; Typis Academiae Carolinae: Stutgardia, Germany, 1788. [CrossRef]

67. Minelli, A. *Understanding Development*; Cambridge University Press: Cambridge, UK, forthcoming.

68. Hendrikse, J.L.; Parsons, T.E.; Hallgrímsson, B. Evolvability as the proper focus of evolutionary developmental biology. *Evol. Dev.* **2007**, *9*, 393–401. [CrossRef]

69. Minelli, A. Evolvability and its evolvability. In *Challenges to Evolutionary Theory: Development, Inheritance and Adaptation*; Huneman, P., Walsh, D., Eds.; Oxford University Press: New York, NY, USA, 2017; pp. 211–238. [CrossRef]

70. Schlosser, G.; Wagner, G.P. Introduction: The modularity concept in developmental and evolutionary biology. In *Modularity in Development and Evolution*; Schlosser, G., Wagner, G.P., Eds.; University of Chicago Press: Chicago, IL, USA, 2004; pp. 1–11.

71. Müller, G.B.; Wagner, G.P. Innovation. In *Keywords and Concepts in Evolutionary Developmental Biology*; Hall, B.K., Olson, W., Eds.; Harvard University Press: Cambridge, MA, USA, 2003; pp. 218–227.

72. Müller, G.B.; Newman, S.A. The innovation triad: An EvoDevo agenda. *J. Exp. Zool. B Mol. Dev. Evol.* **2005**, *304*, 487–503. [CrossRef]

73. Peterson, T.; Müller, G.B. What is evolutionary novelty? Process versus character based definitions. *J. Exp. Zool. B Mol. Dev. Evol.* **2013**, *320B*, 345–350. [CrossRef]

74. Fields, C.; Levin, M. Scale-free biology: Integrating evolutionary and developmental thinking. *BioEssays* **2020**, 1900228. [CrossRef]

75. Kupiec, J.-J. *The Origins of Individuals*; World Scientific: Singapore, 2009. [CrossRef]

76. Griffiths, P.; Stotz, K. Developmental systems theory as a process theory. In *Everything Flows: Towards a Processual Philosophy of Biology*; Nicholson, D.J., Dupré, J., Eds.; Oxford University Press: Oxford, UK, 2018; pp. 225–245. [CrossRef]

77. Griesemer, J. The units of evolutionary transition. *Selection* **2000**, *1*, 67–80. [CrossRef]

78. Dobzhansky, T. Nothing in biology makes sense except in the light of evolution. *Am. Biol. Teach.* **1973**, *35*, 125–129. [CrossRef]

79. Griesemer, J.R. Tools for talking: Human nature, Weismannism and the interpretation of genetic information. In *Are Genes Us? The Social Consequences of the New Genetics*; Cranor, C., Ed.; Rutgers University Press: New Brunswick, NJ, USA, 1994; pp. 69–88.

80. Griesemer, J.R. Individuation of developmental systems: A reproducer perspective. In *Individuation, Process, and Scientific Practices*; Bueno, O., Chen, R.-L., Fagan, M.B., Eds.; Oxford University Press: Oxford, UK, 2018; pp. 137–164. [CrossRef]

81. Minelli, A. Development, an open-ended segment of life. *Biol. Theory* **2011**, *6*, 4–15. [CrossRef]

82. Minelli, A.; Pradeu, T. (Eds.) *Towards a Theory of Development*; Oxford University Press: Oxford, UK, 2014; ISBN 9780191781117.

83. Pradeu, T.; Laplane, L.; Prévot, K.; Hoquet, T.; Reynaud, V.; Fusco, G.; Minelli, A.; Orgogozo, V.; Vervoort, M. Defining "development". *Curr. Top. Dev. Biol.* **2016**, *117*, 171–183. [CrossRef] [PubMed]

84. Bonduriansky, R.; Day, T. Nongenetic inheritance and its evolutionary implications. *Ann. Rev. Ecol. Evol. Syst.* **2009**, *40*, 103–125. [CrossRef]

85. Bošković, A.; Rando, O.J. Transgenerational epigenetic inheritance. *Ann. Rev. Genet.* **2018**, *52*, 21–41. [CrossRef] [PubMed]

86. Richards, E.J. Inherited epigenetic variation - Revisiting soft inheritance. *Nat. Rev. Genet.* **2006**, *7*, 395–401. [CrossRef]

87. Jablonka, E. Epigenetic inheritance and plasticity: The responsive germline. *Prog. Biophys. Mol. Biol.* **2013**, *111*, 99–107. [CrossRef]

88. Jablonka, E. The evolutionary implications of epigenetic inheritance. *Interface Focus* **2017**, *7*, 20160135. [CrossRef]

89. Jablonka, E.; Raz, G. Transgenerational epigenetic inheritance: Prevalence, mechanisms, and implications for the study of heredity and evolution. *Quart. Rev. Biol.* **2009**, *84*, 131–176. [CrossRef]

90. Jablonka, E.; Lamb, M.J. *Evolution in Four Dimensions: Genetic, Epigenetic, Behavioral, and Symbolic Variation in the History of Life*; MIT Press: Cambridge, MA, USA, 2005; ISBN 0262101076.

91. Fusco, G. Evo-devo beyond development: The evolution of life cycles. In *Perspectives on Evolutionary and Developmental Biology*; Fusco, G., Ed.; Padova University Press: Padova, Italy, 2019; pp. 309–318.

92. Oyama, S. *The Ontogeny of Information: Developmental Systems and Evolution*; Cambridge University Press: Cambridge, UK, 1985; ISBN 0521320984.

93. Oyama, S.; Griffiths, P.E.; Gray, R.D. (Eds.) *Cycles of Contingency: Developmental Systems and Evolution*; MIT Press: Cambridge, MA, USA, 2001; ISBN 0262150530.

94. Gorelick, R. Mitosis circumscribes individuals; sex creates new individuals. *Biol. Philos.* **2012**, *27*, 871–890. [CrossRef]

95. Minelli, A. Developmental disparity. In *Towards a Theory of Development*; Minelli, A., Pradeu, T., Eds.; Oxford University Press: Oxford, UK, 2014; pp. 227–245. [CrossRef]

96. Dupré, J. The polygenomic organism. *Sociol. Rev.* **2010**, *58* (Suppl. 1), 19–31. [CrossRef]

97. Bosch, T.C.G.; McFall-Ngai, M.J. Metaorganisms as the new frontier. *Zoology* **2011**, *114*, 185–190. [CrossRef]

98. Gilbert, S.F.; Sapp, J.; Tauber, A.I. A symbiotic view of life: We have never been individuals. *Quart. Rev. Biol.* **2012**, *87*, 325–341. [CrossRef]

99. McFall-Ngai, M.; Hadfield, M.G.; Bosch, T.C.G.; Carey, H.V.; Domazet-Lošo, T.; Douglas, A.E.; Dubilier, N.; Eberl, G.; Fukami, T.; Gilbert, S.F.; et al. Animals in a bacterial world, a new imperative for the life sciences. *Proc. Natl. Acad. Sci. USA* **2013**, *110*, 3229–3236. [CrossRef]

100. Gilbert, S.F.; Epel, D. *Ecological Developmental Biology: The Environmental Regulation of Development, Health, and Evolution*; Sinauer: Sunderland, MA, USA, 2015; ISBN 9781605353449.

101. Wagner, G.P. The biological homology concept. *Annu. Rev. Ecol. Syst.* **1989**, *20*, 51–69. [CrossRef]

102. Wagner, G.P. Homologues, natural kinds and the evolution of modularity. *Am. Zool.* **1996**, *36*, 36–43. [CrossRef]

103. Esteve-Altava, B. Challenges in identifying and interpreting organizational modules in morphology. *J. Morphol.* **2017**, *278*, 960–974. [CrossRef] [PubMed]

104. Callebaut, W.; Rasskin-Gutman, D. (Eds.) *Modularity: Understanding the Development and Evolution of Natural Complex Systems*; The MIT Press: Cambridge, MA, USA, 2005; ISBN 9780262513265.

105. Wagner, G.P.; Pavlicev, M.; Cheverud, J.M. The road to modularity. *Nature Rev. Genet.* **2007**, *8*, 921–931. [CrossRef]

106. Eble, G.J. Morphological modularity and macroevolution. In *Modularity: Understanding the Development and Evolution of Natural Complex Systems*; Callebaut, W., Rasskin-Gutman, D., Eds.; The MIT Press: Cambridge, MA, USA, 2005; pp. 221–238.

107. Wagner, G.P.; Altenberg, L. Complex adaptations and the evolution of evolvability. *Evolution* **1996**, *50*, 967–976. [CrossRef]

108. Klingenberg, C.P. Morphological integration and developmental modularity. *Annu. Rev. Ecol. Evol. Syst.* **2008**, *39*, 115–132. [CrossRef]

109. Klingenberg, C.P. Evolution and development of shape: Integrating quantitative approaches. *Nat. Rev. Genet.* **2010**, *11*, 623–635. [CrossRef] [PubMed]

110. Klingenberg, C.P. Studying morphological integration and modularity at multiple levels: Concepts and analysis. *Philos. Trans. R. Soc. B* **2014**, *369*, 20130249. [CrossRef]

111. Dover, G.; Coen, E. Springcleaning ribosomal DNA: A model for multigene evolution? *Nature* **1981**, *290*, 731–732. [CrossRef]

112. Dover, G. A molecular drive through evolution. *BioScience* **1982**, *32*, 526–533. [CrossRef]

113. Dover, G. Molecular drive: A cohesive mode of species evolution. *Nature* **1982**, *299*, 111–117. [CrossRef]

114. Minelli, A.; Fusco, G. Body segmentation and segment differentiation: The scope for heterochronic change. In *Evolutionary Change and Heterochrony*; McNamara, K.J., Ed.; Wiley: London, UK, 1995; pp. 49–63.

115. Minelli, A. *Perspectives in Animal Phylogeny and Evolution*; Oxford University Press: Oxford, UK, 2009; ISBN 9780198566205.

116. Minelli, A. Tracing homologies in an ever-changing world. *Riv. Estet. N.S.* **2016**, *56*, 40–55. [CrossRef]

117. Minelli, A. *Plant Evolutionary Developmental Biology. The Evolvability of the Phenotype*; Cambridge University Press: Cambridge, UK, 2018; ISBN 1139542362.

118. Robson, G.C. *The Species Problem. An Introduction to the Study of Evolutionary Divergence in Natural Populations*; Oliver and Boyd: Edinburgh, UK, 1928.

119. Bernard, H.M. The species problem in corals. *Nature* **1902**, *65*, 560. [CrossRef]

120. Bessey, C.E. The taxonomic aspect of the species question. *Am. Nat.* **1908**, *42*, 218–222. [CrossRef]

121. Cowles, H.C. An ecological aspect of the conception of species. *Am. Nat.* **1908**, *42*, 265–271. [CrossRef]

122. Calman, W.T. The taxonomic outlook in zoology. *Science* **1930**, *72*, 279–284. [CrossRef]

123. Bakloushinskaya, I.Y. Darwin's heritage: Endless evolution of a species concept. *Russ. J. Dev. Biol.* **2019**, *50*, 287–289. [CrossRef]

124. Mallet, J. Species, concepts of. In *Encyclopedia of Biodiversity*; Levin, S.A., Ed.; Academic Press: Waltham, MA, USA, 2013; Volume 6, pp. 679–691.

125. O'Hara, R. Systematic generalization, historical fate, and the species problem. *Syst. Biol.* **1993**, *42*, 231–246. [CrossRef]

126. Grubb, P.; Groves, C.P.; Dudley, J.P.; Shoshani, J. Living African elephants belong to two species: *Loxodonta africana* (Blumenbach, 1797) and *Loxodonta cyclotis* (Matschie, 1900). *Elephant* **2000**, *2*, 1–4. [CrossRef]

127. International Commission on Zoological Nomenclature. *International Code of Zoological Nomenclature*, 4th ed.; The International Trust for Zoological Nomenclature: London, UK, 1999.

128. Turland, N.J.; Wiersema, J.H.; Barrie, F.R.; Greuter, W.; Hawksworth, D.L.; Herendeen, P.S.; Knapp, S.; Kusber, W.-H.; Li, D.-Z.; Marhold, K.; et al. (Eds.) *International Code of Nomenclature for Algae, Fungi, and Plants (Shenzhen Code) Adopted by the Nineteenth International Botanical Congress Shenzhen, China, July 2017*; Koeltz Botanical Books: Glashütten, Germany, 2018; ISBN 978-3-946583-16-5. [CrossRef]

129. Berendsohn, W.G. The concept of "potential taxa" in databases. *Taxon* **1995**, *44*, 207–212. [CrossRef]

130. Blanc, J. *Loxodonta Africana. The IUCN Red List of Threatened Species 2008*; International Union for Conservation of Nature and Natural Resources: Gland, Switzerland, 2008. [CrossRef]

131. Minelli, A. Taxonomy needs pluralism, but a controlled and manageable one. *Megataxa* **2020**, *1*, 9–18. [CrossRef]

132. Minelli, A. The galaxy of the non-Linnaean nomenclature. *Hist. Philos. Life Sci.* **2019**, *41*, 31. [CrossRef]
133. Lepage, D.; Vaidya, G.; Guralnick, R. Avibase—A database system for managing and organizing taxonomic concepts. *ZooKeys* **2014**, *420*, 117–135. [CrossRef]
134. Lepage, D. Avibase—The World Bird Database. 2019. Available online: http://avibase.bsc-eoc.org (accessed on 30 July 2019).
135. Mishler, B.; Donoghue, M. Species concepts: A case for pluralism. *Syst. Zool.* **1982**, *31*, 491–503. [CrossRef]
136. Kitcher, P. Species. *Philos. Sci.* **1984**, *51*, 308–333. [CrossRef]
137. Ereshefsky, M. Eliminative pluralism. *Philos. Sci.* **1992**, *59*, 671–690. [CrossRef]
138. Ereshefsky, M. *The Poverty of the Linnaean Hierarchy: A Philosophical Study of Biological Taxonomy*; Cambridge University Press: Cambridge, UK, 2001. [CrossRef]
139. Ghiselin, M.T. Species concepts, individuality, and objectivity. *Biol. Philos.* **1987**, *2*, 127–143. [CrossRef]
140. Hull, D.L. Genealogical actors in ecological roles. *Biol. Philos.* **1987**, *2*, 168–184. [CrossRef]
141. Conix, S. Radical pluralism, classificatory norms and the legitimacy of species classifications. *Stud. Hist. Philos. Biol. Biomed. Sci.* **2019**, *73*, 27–34. [CrossRef] [PubMed]

Publisher's Note: MDPI stays neutral with regard to jurisdictional claims in published maps and institutional affiliations.

 *philosophies*

Article

Multiplicity of Research Programs in the Biological Systematics: A Case for Scientific Pluralism

Igor Y. Pavlinov

Research Zoological Museum, Lomonosov Moscow State University, 125009 Moscow, Russia;
igor_pavlinov@zmmu.msu.ru

Received: 10 March 2020; Accepted: 13 April 2020; Published: 15 April 2020

Abstract: Biological diversity (BD) explored by biological systematics is a complex yet organized natural phenomenon and can be partitioned into several aspects, defined naturally with reference to various causal factors structuring biota. These BD aspects are studied by particular research programs based on specific taxonomic theories (TTs). They provide, in total, a framework for comprehending the structure of biological systematics and its multi-aspect relations to other fields of biology. General principles of individualizing BD aspects and construing TTs as quasi-axiomatics are briefly considered. It is stressed that each TT is characterized by a specific combination of interrelated ontological and epistemological premises most adequate to the BD aspect a TT deals with. The following contemporary research programs in systematics are recognized and characterized in brief: phenetic, rational (with several subprograms), numerical, typological (with several subprograms), biosystematic, biomorphic, phylogenetic (with several subprograms), and evo-devo. From a scientific pluralism perspective, all of these research programs, if related to naturally defined particular BD aspects, are of the same biological and scientific significance. They elaborate "locally" natural classifications that can be united by a generalized faceted classification.

Keywords: research programs; scientific pluralism; taxonomic theory; taxonomic pluralisms; typology; phylogenetics; biosystematics; numerical taxonomy; biomorphics; evo-devo

---

1. Introduction: Monism vs. Pluralism in Biological Systematics

The dilemma of scientific monism vs. scientific pluralism arose simultaneously with the beginning of modern science development. It can be represented, in a brief and simplified form, as follows [1,2]. In the first case, it is presumed that, both in science in general and in any science branch, there might and should be the only one actually scientific approach providing the only one "right" theory or concept most adequately describing the cognizable world. In the second case, it is presumed that any natural phenomenon, be it entire nature or any of her manifestations, is too complex to be embraced by only one approach and respective theory/concept. Correspondingly, such a "patched" phenomenon might and should be described by several partial theories/concepts, with each capturing its particular manifestation. Therefore, it is their combination that provides an integrated representation of the phenomenon in question. Respectively, the science appeared to be "patched" by research programs, with each developing a particular approach most relevant to a certain manifestation of the respective phenomenon being perceived [3]. Acknowledging such partitioning of both the natural phenomena and scientific activities is of fundamental importance for understanding how various disciplinary fields are individualized and interrelated.

In the life sciences, biological systematics (or systematic biology) plays a fundamental role in such partitioning of biological matter. It studies taxonomic diversity of living beings, as one of the manifestations of the total biological diversity (a.k.a. biodiversity, in its widely adopted scientific sense), which was defined by specific relations between organisms expressed in and revealed by analysis of the diversity of their proper ("inner") features. Other manifestations of biodiversity, defined by relations of organisms with their environments, are studied by other biological disciplines (biogeography, synecology, sociobiology, etc.). The main function of systematics is to reveal and describe the structure of taxonomic diversity and, thus, to shape, by and large, the subject areas for many other biological disciplines. Due to this, it attracts the attention of both philosophers and biologists exploring the foundations of biology and some of its key concepts such as evolution, hierarchy, species, etc. [4–11].

Taxonomic diversity itself is a multifaceted natural phenomenon because of the multiplicity of relations between organisms shaping it and organismal features expressing these relations. Therefore, systematics develops a variety of approaches for studying the various manifestations of taxonomic diversity. The most important are formalized as taxonomic theories and developed into research programs peculiar to this discipline. They differ in their ontological and epistemological foundations, in their principles of defining objects and tasks, in the methods of exploration, and in the modes of representation of the structure of biodiversity by classifications. These programs change with the development of systematics and biology, depending largely on changes of general scientific-philosophical contexts. In its turn, the changes of the programs dominating at one or another stage in the history of systematics have a significant impact on understanding how biodiversity is structured and, accordingly, what the structure of the entire subject area of biology is.

Since systematics deals with the same natural phenomenon (the above-mentioned taxonomic diversity), a principal question arises, namely, whether it has to follow the only universal research program or whether there can be several programs equally viable. These two positions, which are of a rather philosophical nature, are known as taxonomic monism and taxonomic pluralism, respectively, and they have been actively discussed over the last several decades [2,8,11].

Taxonomic monism presumes that taxonomic diversity should be considered within a unified conceptual approach and described by a single "right" classification, be it either the Natural system of earlier classics (Linnaeus), the general-purpose classification of phenetics (Gilmour), or the phylogenetic classification of phylogenetics (Hennig). The most devoted adherents of each of these approaches, being monists, believe that it is their theory that says the "final word" in biological systematics and should be accepted as such by the entire taxonomic (and eventually biological) community.

On the contrary, taxonomic pluralism acknowledges that taxonomic diversity is multifold and that all its manifestations are of more or less equal biological meaning and scientific (cognitive) status. Respectively, neither taxonomic theories and research programs dealing with such particular manifestations may pretend to gain a privileged position in biological systematics. Each is significant by their careful examination of particular manifestations of taxonomic diversity. For instance, phylogenetic program structures taxonomic diversity historically, while typological program does that structurally, and biomorphic program uncovers complex morpho-ecological aspects of taxonomic diversity.

The first attempts to recognize explicitly and to discuss the research programs in systematics, as well as to evaluate their scientific status, were undertaken in the 1960s–1970s. There appeared to be acknowledged but three principal "systematic philosophies" most vividly discussed at that time, namely, phenetics, cladistics, and evolutionary taxonomy [5,6,12–14]. However, this "three philosophies" viewpoint did not take into consideration or drastically reduce the significance of other "philosophies" that were not so actively discussed then (typology, biosystematics, ecomorphological approach, etc.). Therefore, such an oversimplification provided a very distorted representation of the theoretical foundations of biological systematics, including diversity of the research programs actually operating in it, their historical and scientific-philosophical roots, their mutual interactions and influences, and their contributions toward the development of both systematics itself and biology overall.

In this article, a brief overview of the principal research programs in biological systematics is provided, guided by their scientific and philosophical foundations, regardless of their general acceptance [11]. One of the main tasks is to show why and how these research programs in biological systematics are developed, and why it is normal for the latter to be pluralistic in this respect. Therefore, attention is paid toward ontological and epistemological prerequisites for elaborating taxonomic theories and for developing the research programs implementing them. The next key task is to show the real diversity of these programs, contrary to a received viewpoint reducing it to a minimal level. All these tasks explain a pretty extensive list of references, even though it only includes the most important issues.

In accordance with the above tasks, the present overview begins with the illumination of some basic ideas concerning the philosophical foundations of biological systematics, as they are understood by the author. This includes also consideration of some general principles of construing taxonomic theory.

2. Some Basic Elements of the Systematic Philosophy

Any general philosophy of science deals primarily with the justification of theoretical knowledge in science. Thus, the "philosophy of systematics" actually presumes analysis of the possible ways to develop the theoretical foundations of biological systematics as a whole.

If systematics is considered a scientifically sound discipline, it is to develop its own taxonomic theory (TT). However, despite a significant number of books, in which principles or foundations of systematics are considered, no sufficiently well-substantiated TT has been elaborated so far. Moreover, hardly any satisfactory understanding seems to exist among taxonomists as to what kind of theory it should or could be. There were only a few earlier attempts to consider some basic premises and principles of what might be called the beginnings of TT, but they were too formal to be biologically sound [7,15,16]. The main reason for such a very deplorable situation is that the substantiations of particular classification approaches have been predominated previously, instead of developing the TT in its general understanding, which would cover the entirety of biological systematics. Such a theory should deal with the multiplicity of both the manifestations of biodiversity and the ways to study it.

Any serious consideration of this important issue would first require an ascertainment of how theories of various levels of generality can be built in different scientific disciplines. However, this would lead us away from the main topic of the present paper, especially when taking into consideration the diversity of the viewpoints on this matter. Therefore, without going deep into this issue, it seems to be enough, for the purposes of this article, to offer the following general declarations concerning just the taxonomic theory [11,17]. The latter is defined as a conceptual system containing generalized theoretical knowledge about what and how biological systematics explores. The answers to the question "what?" make up the ontological part of the theory: they fix the essential characteristics of the object of systematic research. The answers to the question "how?" constitute the epistemological part of the theory: they fix the basic principles of the study of this object. Together, these answers constitute a conceptual framework for defining a TT.

It follows that the main purpose of any TT is framing the theoretical context in which systematic research is conducted and concrete classifications are developed. As such, this theory serves as a general basis for the formation and functioning of any research program in biological systematics. Development of such a theory is the most important and an ultimate aim of taxonomy as a theoretical part of biological systematics.

2.1. One Umgebung—Many Umwelts

One of the first tasks of TT, in its most general sense, is defining an object (or a subject area) of the entirety of biological systematics. Many serious problems arise related not even to the theory itself, but rather to the philosophy of systematics—not in the above Hull's [13], but in a more general sense, concerning its ontological foundations. They deal with illumination of what this biological discipline and its various sections investigate.

All natural sciences, by an initial condition, are aimed ultimately at comprehending and describing nature in its entirety. However, since nature is so global and diverse in all its manifestations, while the human cognitive possibilities are so limited, it is fundamentally impossible to embrace it as a whole with a single gaze and to reach a complete knowledge of nature expressed by a single exhaustive generalization. Instead of a "global" comprehension of nature, it is its "local" manifestations (particular aspects, fragments, etc.) that are actually being explored.

According to the contemporary conceptualism, with Quine's concept of ontological relativity constituting one of its cornerstones [18,19], such fragmentation of nature is shaped by a knowing subject (be it either a person or a scientific community) depending on the latter's cognitive tasks. In this connection, distinguishing the two main levels in reality suggested by the Baltic German zoopsychologist von Uexküll [20,21] seems to be very attractive and thought-provoking. According to his idea, the whole of nature as such (Umgebung) and its actually perceived manifestations (Umwelts) are to be distinguished in a cognitive situation dealt with by natural science. Although Uexküll himself meant selective "biological" perceptions of the environments by particular organisms, his rather metaphoric concept may be widened to include human cognitive activity, which is no less selective with respect to nature being perceived [22].

Thus, from a conceptualist perspective, each particular cognitive Umwelt is fixed by a knowing subject in the course of the latter's cognitive activity to shape a kind of conceptual reality not existing out of this cognitive and perceptual activity. As such, it represents a particular cognizable reality, and multiplicity of these Umwelts implies multiplicity of the approaches dealing with them. Therefore, any cognitive activity begins with preliminarily outlining a particular Umwelt and continues with exploring and describing its properties. In the case of biological systematics, taxonomic diversity can be treated as its Umgebung, while its various manifestations are its various respective Umwelts.

Since systematics deals with taxonomic diversity as a natural phenomenon, its particular Umwelts are to be distinguished as naturally as possible. The best way to ensure this seems to be to do so by indicating the supposed objective causes (initial, proximal, formal, material, etc.) structuring the biota, and, thus, generating various manifestations of diversity of organisms. It is presumed that a cognitive Umwelt thus outlined corresponds to a certain biological phenomenon, which makes its exploration biologically meaningful. On the contrary, an Umwelt fixed by a formal ontology, i.e., without reference to certain natural causes, is apparently devoid of biological meaning.

Different cognitive Umwelts corresponding to particular manifestations of taxonomic diversity are shaped by certain concepts, each defining the properties most significant for individualizing the respective Umwelts. For example, this may be diversity of archetypes or biomorphs, or the hierarchy of monophyletic groups, etc. The phylogeny that generates the latter hierarchy can be interpreted classically in its wholeness (according to Haeckel or Simpson) or reduced to cladogenesis (according to Hennig). On the contrary, one can discard all prior considerations of the structure and causes of taxonomic diversity and deal with separate physically perceived organisms. This also requires specific background knowledge, though, with very poor ontology greatly trimmed by means of the "Occam's razor".

As is evident from the preceding, any transition from the whole Umgebung to one or another particular Umwelt is based on a reduction operation, as far as any isolation of a part from its whole means reduction of the latter. In systematics, such a reduction begins with "cutting" something called biodiversity from the entirety of nature. At the next reduction step, taxonomic diversity is singled out from the entirety of biodiversity, with the latter's other manifestations (ecological, biogeographic, biosocial, etc.) being discarded. Then taxonomic diversity undergoes further decomposition by distinguishing its own aspects, e.g., phylogenetic (emphasis on kinship), typological (emphasis on structural plans), ecogenetic (emphasis on diversity of populations), etc. Furthermore, within phylogenetically defined diversity, its cladogenetic and anagenetic aspects can be distinguished and analyzed separately.

Thus, the above transition from overall Umgebung to particular Umwelts can be represented as a kind of reduction cascade at different steps of which particular exploratory tasks of different levels of generality are successively formulated and solved. All of this is accomplished by a knowing subject (as specified above) with the help of various epistemic tools. It is this subject that decides what is significant and what is not for distinguishing particular Umwelts and extracting them from the Umgebung. Evidently, such a stepwise reduction leads to an unavoidable sequential loss of some part of the objective content of the entire Umgebung at each reduction step. Therefore, an ontology defined by this content is the richest at the very beginning of the reduction cascade and most poor at its end. Accordingly, in the same direction, an accumulated effect of a subjective "input" into particular conceptually construed Umwelts increases. This issue seems to be of prime importance for understanding of "naturalness" (in the above sense) of particular manifestations of taxonomic diversity and, respectively, biological meaningfulness of different taxonomic theories and research programs studying them (see the next section).

It is also important to stress that such a reduction is potentially multiplied at every step of the cascade. This is because taxonomic diversity as a complexly organized natural phenomenon can be represented by several more simple cognitive models (in their general epistemic sense from Reference [23]). Such a multiplicity of the Umwelts, recognized at each step of the reduction, is an ontological prerequisite of an increase of taxonomic plurality descending from top to bottom of the reduction cascade.

2.2. Taxonomic Theory as a Quasi-Axiomatics

Every natural science theory includes some elements of axiomatics. This means that it contains more or less clearly formulated statements about the subject being studied (analogues of axioms) and the principles of its research (analogues of inference rules). Considered from a philosophical standpoint, the former constitute ontology, and the latter constitute epistemology. Such a construction of a taxonomic theory (TT) by using at least some elements of the axiomatic method is advantageous in that it allows us to formulate its basic statements more explicitly, and, thus, to structure the theory itself.

Attempts of this kind were undertaken repeatedly. Some were aimed at developing universal mostly formal systems [7,15,16], while others dealt with the foundations of particular research programs (phenetic, phylogenetic, etc., i.e., [24–27]). Here, the author's position is presented very briefly to show why and how taxonomic theories underlying research programs in biological systematics can be structured and justified [11,17]. The main concern of this section is not to suggest a version of the TT but rather to consider some general principles of its development.

It must be emphasized first that a TT is to be developed as a quasi-axiomatics. This means that, unlike formal axiomatic systems of mathematics and logic, its basic conceptual constructs are initially introduced as biologically sound. This is provided by direct reference to a certain objective (real) manifestation of biodiversity (i.e., typological or phylogenetic pattern), which establishes desirable correspondence between an Umwelt (naturally defined, see previous section) and respective set of quasi-axioms, which makes the latter biologically meaningful. It is quite important to stress that the same meaning can be ascribed to classifications based on these quasi-axioms, while formal axioms with no reference to an Umwelt make respective classifications also formal (biologically meaningless).

Usually, in various systematic textbooks, all such theoretical premises are called principles without distinguishing between their different cognitive functions. However, as seen from the example above, in the framework of the axiomatic approach, it is necessary to divide them into two main categories, namely, quasi-axioms and inference rules. The former have an ontological status and outline an Umwelt under study, while the latter have an epistemological status and deal with the principles of investigation of this Umwelt. Only these inference rules seem to deserve being termed "taxonomic principles".

Though an axiomatic method of construing any theory presumes strict and unambiguous definitions, this requirement cannot be followed literally in the case of natural science disciplines including biological systematics. Its implementation is limited by the principle of an inverse relationship between the rigor and the meaningfulness of the concept definition [28]. The more strictly a concept is defined, the less likely there is something in nature to which it may correspond [29]. Therefore, any definition of an Unmwelt, claimed to be biologically meaningful, is deemed to be imprecise semantically and should be formulated by taking into consideration some conditions of the fuzzy logic (as it is defined by Kosko [30]). The latter means, among others, that such "fuzzy" definitions seem to entail an unfeasibility of their strict and unequivocal applications in studying the structure of biodiversity. The biological concepts that come to mind as most pertinent to this issue are those of taxonomic rank [8,11,31], homology [32–35], and, of course, species [8–10,36–40]. The impossibility of their strict and unambiguous definitions leads to the fuzziness of these concepts, which is reflected by a plurality of their particular meanings.

It is to be stressed especially that quasi-axioms and inference rules within a TT do not work separately, but conjointly in a single package. In general, this condition is formalized by the principle of onto-epistemic correspondence, which means that the basic (for the given TT) statements relating to ontology and epistemology should be meaningfully compatible with each other [11,17]. For example, if an Umwelt is defined phylogenetically, then the principles specify how a classification should be developed to reflect the phylogenetic pattern only.

It follows from the above stepwise reduction cascade (see Section 2.1) that the sequential reduction of the overall Umgebung to a certain set of Umwelts results in the generation of respective particular quasi-axiomatics of different levels of generality. It is presumed that each set of quasi-axioms outlines a particular Umwelts of a certain level of generality. On this basis, a hierarchy of TTs allocated to these levels can be consequently construed. Thus, taxonomic theory considered in its most general sense is a rather complex multilayer construct. It consists of several levels of theoretical generalizations with each solving specific tasks allocated to a respective level of a reduction cascade. Theoretical provisions of the highest level of generality constitute the general taxonomic theory (GTT), while those belonging to the lower levels are particular taxonomic theories (PTTs).

In this hierarchical conceptual pyramid, the GTT plays the role of a framework concept in relation to various PTTs and can be considered a taxonomic meta theory (i.e., theory of theories) for them. Within such a pyramid, particular PTTs arise as different details of the GTT statements. The main task of the latter is to outline correctly (including being biologically sound) the cognitive situation for the entirety of biological systematics, including its basic ontological and epistemological components. Thus, GTT can be imagined as a set of interconnected general judgments about the general properties of taxonomic diversity (ontology) and the general principles of its study (epistemology). This theory is intended not to elaborate concrete classifications, but rather to formulate (as prescriptions and restrictions) the grounds for possible ways to formulate and solve the exploratory tasks that systematics deals with. Thus, it is the GTT that can more than justifiably be claimed as a systematic philosophy.

There are two principal modes of construing the GTT. One of them refers primarily to ontological quasi-axioms from which certain principles are deduced, while another accentuates epistemological reference rules (principles) equally applied to any natural phenomena. Thus, taxonomic pluralism is observed even at the most general level of the theoretical basis of biological systematics.

The ontology-based GTT specifies first how a particular Umwelt is to be outlined. For instance, it specifies whether causes of the diversity of organisms should be indicated or not, and if indicated, which particular ones—historical in phylogenetics, structural in typology, or functional in biomorphic, etc. If the evolutionary process is referred to as the main cause of taxonomic diversity, it can be interpreted as an adaptatiogenesis or as a more "formal" cladogenesis. On this basis, it is then specified (quasi-axiomatized) which particular relations between organisms are taken into account—only kinship, only similarity, or some combinations thereof. Based on these basic assumptions, certain taxonomic principles are developed. Some of them entail homologization, character weighting, similarity assessment, etc., while others deal with inferring particular phylogenetic schemes, and the next deal with elaborating phylogenetic classifications based on these schemes. The same general design is true for any other biologically meaningful quasi-axiomatics, be it typological, biomorphic, or otherwise.

On the other hand, the epistemology-based GTT presumes that any Umwelts, however defined, are not specific with respect to their properties (ontology), so the main task is to elaborate certain universal principles of their analysis (epistemology) independent of particular ontologies. The latter means that systematic research should be subordinated to certain fairly formalized universal reference rules. Particular implementations of such accentuation are PTTs in which logical or computational procedures are set as being of primary importance (see below).

From a formal standpoint, any properly construed quasi-axiomatic systems relevant to the task systematics deals with have an equal cognitive potential. However, as stressed above, as far as systematics is a biological discipline exploring biodiversity as a real phenomenon, evaluating the research potentials of all possible TTs is to be based on their biological meaningfulness. The latter means, first, that their basic quasi-axioms should refer to certain natural biological phenomena, and, second, the respective Umwelts representing the latter should be as least reductionist as possible. Therefore, ontology-based TTs seem to be more significant as compared to epistemology-based TTs, and, among the former, those with a rich ontology are more significant.

As seen, the outlined quasi-axiomatic method of developing the theoretical foundations of biological systematics makes it rather easy to understand the whole structure of both the GTT (which is still an imaginary, rather than a well-established, construct) and the reasons of plurality of PTTs detailing the latter.

3. An Overview of the Research Programs in Systematics

Research programs in biological systematics appear and function as a means of implementation of particular taxonomic theories. The latter develop not by themselves, but in a certain philosophical–scientific context, with one way or another responding to the challenges that systematics faces at one or another stage of conceptual development of natural science.

The subsequent sections provide a review of the research programs in biological systematics that have developed over the 20th and at the beginning of the 21st centuries. Some of them basically continued the ideas formed in the 19th century (typological, phylogenetic, etc.) while others emerged de novo in this period (phenetic, numerical, evo-devo, etc.). It is to be stressed that this review is based on preceding considerations of the principles of construing and evaluating the corresponding taxonomic theories. Accordingly, in characterizing the latter, the most focus is put primarily onto their philosophical (ontological and epistemological) foundations, regardless of the popularity they enjoy among biologists at various stages of the development of biological systematics. For this, the order of presentation of these programs corresponds basically to a certain scale of the richness of their ontologies. The below account opens the most reductionist programs (phenetic, rational, numerical) and closes with the biologically soundest programs (typological, biomorphic, phylogenetic, evo-devo).

3.1. The Phenetic Program

This program develops and formalizes an old idea of empirical knowledge and, as such, goes back to folk systematics (see References [11,41,42] on the latter). The beginning of its persistent formation in scientific systematics was laid down by the works of the anti-scholastics of the second half of the 18th century. Usually, the French naturalist M. Adanson's approach is mentioned in this connection—to the extent that the founders of modern numerical phenetics used to call it "neo-Adansonian" [24,25]. However, such identification was shown to be incorrect [5,43], as the Adansonian methodology actually foreruns one of the numerical cladistic approaches [44]. The genuine phenetic concept was expressed at that time by the German naturalist J. Blumenbach. He stated in his "Handbuch der Naturgeschichte" that the "animals that are similar in 19 structures and differ only in the twentieth should be grouped together" (see Reference [11]).

In the 20th century, the phenetic program in its rather strict sense was substantiated by a classification theory based on the positivist philosophy of science, as its early ideologists stated explicitly [24,25,45]. It is closely related to the numerical program (see Section 3.3), so they are often considered conjointly. However, this is not correct. The phenetic theory deals with what is studied (ontology), while the numeric one deals with how that "what" is studied (epistemology).

According to this philosophy, in a cognitive situation of phenetic systematics, the background knowledge is minimized in order to exclude its "metaphysical" content (such a reference to evolution).

Respectively, a phenetically defined Umwelt is simply a set of observable physical bodies with their characters, i.e., organisms. At the same time, the subjective influence is also excluded as much as possible. All operations on those "physical bodies"—their description, comparison, etc.—should be depersonalized and reduced to some elementary automated actions. Phenetic classification is developed as purely empirical (in a philosophical sense). It should be nothing more than a generalization of the observed facts, which makes it independent of any biological theories.

In developing phenetic classifications, the only basis for grouping organisms is their mutual similarity as such, which is assumed to be both objective and theory-neutral. However, both these presumptions are not true, as any similar relation is established by a knowing subject depending on the conceptually framed exploratory tasks [46–49]. Thus, the phenetic program seems to lack its initial "as-if empiric" philosophical background. This similarity is evaluated across the totality of the unit (elementary) characters used in the comparisons without any preliminary assessment of their significance ("weight"). One of the most serious restrictions on the choice of characters is that they should describe organisms themselves (morphology, physiology, genetics, etc.), but not their relations to the environment (ecology, etc.) [24,25]. Such a "weighting" ascends the essentialism of earlier (scholastic) taxonomic theory [11]. The resulting taxa in phenetic classifications are designated as phenons [46–49]. Ontological interpretation of both them and their ranks is nominalistic.

The main purpose of phenetic classification is quite pragmatic. It should not reflect some mysterious "naturalness", but instead should be "useful". The usefulness of a classification depends on its informativeness, i.e., on the volume of information about the diversity of the organisms contained within it. Maximizing the information content in the classification is achieved by increasing the number of characters used for its elaboration. Roughly speaking, the more characters, the better. This condition is substantiated by the positivist principle of total evidence coupled with an ad hoc hypothesis of character non-specificity and the mathematical principle of convergence. It is assumed that classifications, if starting from different initial sets of characters, should converge asymptotically with the maximum possible increase of the number of characters [24,25].

The purely technical nature of phenetic classifications means that they are not evaluated from the point of view of their naturalness in its traditional meaning. Instead, it is replaced by certain operational criteria of the classification informativeness assessment (Gilmour-naturalness). The most informative classifications that can be used to solve many different tasks are termed general purpose ones. With reference to the above principle of convergence, the potential attainability of a single stable reference system as an ideal of phenetic systematics is supposed. Along with it, various special purpose classifications can and should be elaborated to solve certain particular research and applied tasks. There can be a lot of them and they can be very different in their information content, but all of them are subordinate with respect to the general-purpose classification. Thus, the phenetic program is monistic with respect to the general-purpose classification and pluralistic with respect to the special-purpose ones.

The phenetic program, supplied with the numerical methods, was most popular in the 1960s. At present, it has been supplanted by the phylogenetic program in its cladistic and molecular versions (see Section 3.7). In this regard, it is important to keep in mind that the latter borrowed some important points of the phenetic theory. Thus, in molecular phylogenetics, an idea of the reduction of an organism to a set of automatically identified unit characters appeared to be perfected: these are nucleotide base pairs in DNA and RNA sequences. In cladistics, clear elements of the phenetic theory are introduced by the above mentioned principle of total evidence according to which, roughly speaking, "the more characters, the better" [50–52].

From the point of view of the philosophy of science that focuses on modern conceptualism (in the above sense of Reference [18]), the main problem of the positivism based phenetic program is that a strictly empirical knowledge, devoid of any theoretical basis referring to a certain natural ontology, is impossible in the natural science [18,53]. This key standpoint means that both the phenetic theory itself and the program implementing it dropped out of the framework of contemporary science with its dominating post-positivist philosophy.

3.2. The Rational Program

Rationality, understood in its general sense, constitutes the very basis of the science distinguishing it from other forms of the cognitive activity. Among various versions of scientific rationality [54,55], a deductive one is the most relevant for substantiation of the rational program in systematics, as it is understood by the author [11,56]. It is based on acknowledging the paramount importance of such syllogistics in which particular judgments are deduced from general ones. It is presumed that the truth of the former entails the truth of the latter. The general aim of such an approach in biological systematics is to develop a kind of rational classification, which may be classified as "all elements of which are derived on the basis of some general principles, certain theory" [57]. It is clear that formulation of taxonomic theory as a kind of quasi-axiomatics (see Section 2.2) fits completely the conditions of such rationality.

Initial judgments, with which formation of the rational program in systematics began, are two-fold. Some of them are related to the object being researched, i.e., to ontology: this can be termed as an ontic rationality. Others relate to the principles of research, i.e., to epistemology: accordingly, they constitute an epistemic rationality. Thus, this rational program is divided into two subprograms, which are called onto-rational and episto-rational ones. They are similar in the deductive (with reference to the concepts of higher levels of generality) substantiation of particular judgments but differ in the content of the general ones [11,56].

One of the first versions of the onto-rational subprogram in systematics was proposed by the Swiss botanist A.-P. de Candolle (actually, he was probably the first to coin the term "rational classification" in biology) at the beginning of the 19th century, who based his theory on the principles of symmetry borrowed from crystallography [58,59]. In a more general form, the idea of rational systematics was formulated 100 years later by the German natural philosopher H. Driesch [60]. He called for uncovering some general law of orderliness of diversity of biological forms that would be analogous to the natural laws of orderliness of chemical elements in physical chemistry or geometric figures in geometry. A rational classification based on such a law is presumed to become a powerful heuristic that allows certain predictions about still unknown forms. Thus, it is reasonable to call the onto-rational systematics nomothetic [61]. It reveals the general laws of taxonomic diversity instead of presenting the latter as a list of taxa with their diagnostic characters. It is evident that the taxa recognized within the onto-rational taxonomic theory are interpreted realistically as natural kinds in the sense of W. Quine and others [62,63].

Implementation of one of Driesch's ideas led to an aspiration to elaborate parametric classifications of living forms analogous to the above periodic system of elements in chemistry [64,65]. The latter means that such classifications should not be construed on a strictly hierarchical (vertical) basis, but rather horizontal relations between biological forms being most important. The main problem in this case is that the biological forms are much more complex than the chemical elements. Therefore, it is difficult to recognize a key parameter by which their periodic system could be arranged. Such a parameter is usually suggested to define the organismal complexity, but it does not lend itself to a universal, satisfactory enough definition, which allows for the development of a single scale of complexity [66,67]. Thus, the principal idea of this version of the onto-rational program in systematics seems hardly resolvable in general. However, some of its applications are of certain interest as they may uncover some important properties in the ordered structure of biodiversity.

There is another partial taxonomic theory called rational by its creators [68–70], which fits the conditions of onto-rationality. In this case, rationality presumes deducing classifications from the orderliness of the diversity of ontogenetic patterns. In this case, this version is considered a part of the epigenetic typological subprogram (see Section 3.4).

The episto-rational subprogram focuses on certain general inference rules governing the particular exploratory procedures in systematics. Such a standpoint presumes that it is these rules that are primary, and all taxonomic principles and methods are to be inferred from them, while ontological considerations are minimized. In the classical systematics of the 16th to 18th centuries, this general conception led to a fundamental monistic idea of the natural method, a proper application of which would provide the natural classification. According to M. Adanson [71], this method "should be universal or overall, i.e., there should be no exception for it". This general idea was first implemented by scholastic systematics of the 16th to the 18th centuries in a form of the universal genus-species scheme [11]. Post-scholastic systematics rejected this particular scheme, but the very idea was retained and led eventually to the development of two research subprograms. In one of these, the logical argumentation is taken as the basis, while another accentuates on mathematically based judgments.

An idea of strict subordination of taxonomic principles of biological systematics to some general logic was expressed in recent times by several authors [15,16,57,72]. As a matter of fact, it explicates a rather old idea that any classification is but a logical procedure. Some fragmentary attempts to implement it were based on application of the requirements of the formal axiomatic systems (see Section 2.2). A more recent and quite developed general solution is offered by classiology [73]. Ideologists of the "logical systematics" presume that any classification theory derived from some "general logic" is applicable to any phenomena (natural, social, etc.) studied by any classifying sciences, regardless of their natural ontology. However, no one pure logical taxonomic theory for biological systematics has been proposed so far, and no biological classifications of such a kind are known. Thus, the main idea of a would-be logical research program in systematics still remains only a kind of "declaration of intent".

In considering the productivity of such a theory for systematics, of prime importance is the fundamental issue whether its basic idea has any relevance to the tasks this biological discipline deals with. As a matter of fact, systematics asks a question about objective (real) biodiversity and tries to answer it in a sound manner. However, every logical system is merely a specific tool designed to ensure the logical consistency of derivative particular statements with respect to the more general ones, all within a particular formal axiomatic system. Such a tool asserts the logical truth of the conclusions with respect to that system, but it does not say anything about their natural truth with respect to the reality being studied because it does not consider this respect at all [74]. Thus, the program in question does not seem to expect either the very question or an answer as to how "logical" classifications may relate to the reality and, if they relate, how to ascertain this.

If the above declaration is followed literally, it is to be taken into consideration that any appeal to some general logic looks very naïve, as there exist many formal logical systems [74,75], including some (binary, probabilistic, fuzzy, modal, etc.) that are relevant to systematics. Thus, the next key question arises—now it is about the basis for a choice of particular logical systems for deducing particular taxonomic theories applicable for elaborating biologically meaningful classifications.

As far as such a question is concerned, one of the possible answers is provided by the above principle of onto-epistemic correspondence (see Section 2.2). According to the modern conceptualist standpoint [18] underlying this principle, it is ontology that drives epistemology (including logical inference rules), and not vice versa. In other words, the principles of elaboration of biologically meaningful classifications are to be inferred from background assumptions about properties of taxonomic diversity rather than from any pure logic [11,17,76]. Thus, it becomes clear from this standpoint that any kind of a pure logical research program in biological systematics seems to be unfeasible.

A part of the episto-rational subprogram is the numerical one based primarily on the mathematical foundations of taxonomic principles and methods. Because of its significant influence on the development of biological systematics in the 20th century, it likely deserves the status of a research program of its own, so it is considered in the following subsection.

3.3. The Numerical Program

As noted, this is actually one of the versions of a wider episto-rational subprogram, though deserving a status of a distinct program. Its source lies in the natural philosophical idea that the "book of nature is written in the language of mathematics" announced by the Italian physicist G. Galilei in the 17th century [77]. Based on this, the Prussian philosopher I. Kant at the end of the 18th century expressed one of the key ideas of modern physicalism: "in any special doctrine of nature, there can be only as much proper science as there is mathematics therein" [78].

The 19th century English naturalist H. Strickland should likely be considered one of the forerunners of the numerical program. Based on the then rather popular so called "taxonomic map" metaphor [79], he likened the similarity between groups of organisms to the distance between territories on a geographical map: the longer the distance, the less the similarity [80]. At the beginning of the 20th century, the Russian biologist E. Smirnov put the key idea of this program this way: you need to "establish those rules and laws that determine the relative position of the phenomena studied. The expression of these laws in the form of mathematical formulas is the highest goal that systematization strives for" [81]. Smirnov called such a taxonomy "exact". In the future, its supporters called it "numerical" and then, quite straightforwardly, "mathematical". The general design of the numerical program in taxonomy was framed mostly in the 1960s–1970s [11,82]. Based on the above Kant's aphorism, adherents of this program consider it the only one deserving the title of scientific in the physicalist sense of the latter and deem it as the most significant scientific revolution in contemporary systematics.

The main content of this program can be formulated in two general principles. First, relations (similarity, kinship, etc.) between organisms and sets thereof can and should be measured quantitatively. Second, the structure of relations thus quantified can and should be transformed into a classification by means of quantitative methods. Implementation of this program is provided by many methods developed to solve various particular classificatory tasks within the general numerical idea.

It is clear that the numerical program, based mainly on epistemology, does not have any subject domain of its own. It concerns the issues related to ontology as far as an Umwelt to be studied is to be adapted to the needs of quantitative methods by means of its significant reduction. For instance, the organisms are reduced to a matrix of an uncorrelated formalized character. Thus, any function of this program is limited to serve as an analytical supplement to research programs based on ontological considerations—only those that consider such a reduction is acceptable.

Depending on particular biologically meaningful tasks solved using quantitative methods, the program under consideration is divided into two main subprograms: viz. numerical phenetic and numerical phyletic ones.

Numerical phenetics [24,25,83,84] provides quantitative methods for implementation of the phenetic and, partly, the bio-systematic programs (see Sections 3.1 and 3.6 on them). In this case, a classification procedure is based on quantitative assessments of the similarities as such, and character weighting is minimized because of the lack of any background knowledge underlying it. The main task is to produce such a pattern of similarity relations among phenons in which the differences within each of them are minimized and the differences between them are maximized with this pattern being transformed subsequently into a phenetic classification.

Numerical phyletics [83–87] develops quantitative methods for implementing the phylogenetic program (see Section 3.7 on it). Its methods are designed to facilitate the reconstruction of phylogenetic relationships. Accordingly, similarity is considered as an indicator of kinship, and characters are weighed to select the most reliable indirect evidence of kinship. Construction of phylogenetic trees is narrowed down to a formal graph theory without any biological considerations [88]. An ultimate goal of numerical phyletics is to develop a tree-like structure that can be interpreted as a network of kinship relationships among supposedly monophyletic groups, and, thus, is capable of being transformed into a phylogenetic classification.

Theoreticians of the numerical program, based on their scientific and philosophical preferences, see its undoubted advantages in objectivity, formalization, repeatability, and exactness of the classification techniques. However, the latter hold initially one fundamental limitation provided by axiomatic justification of pure mathematical methods. The point is that any formal axiomatics contains a certain element of subjectivity [89,90], so an objectivity attributed to both the methods developed on its basis and the results of their application, considered philosophically, is nothing more than a myth, although very widespread. Actually, it is eligible to discuss whether a method is true or false in the logical sense only with respect to the axiomatics underlying it, but not with respect to a particular Umwelt analyzed with it. Thus, it is hardly possible at all to say if a classification obtained by a pure formal method is true or false as a cognitive model (in the sense of Wartofsky, see Reference [23]) of this Umwelt, which makes no sense in any consideration whether such a classification is objective or not [4]. The only point that can be discussed rightfully is intersubjectivity [53,91], which means that different researchers, solving the same standard problem with the same standard method, get the same standard result (notorious repeatability). As for the exactness of the axiomatically substantiated methods and the results of their application, it is determined only within the framework of the formalizations embedded in the initial axiomatics, and not necessarily so in others [89,90].

Opponents of this program consider its main idea—primacy of the formal method over biological content—flawed. It reduces biologically meaningful tasks to purely technical ones and, thus, from a metaphysical perspective, "puts the cart in front of the horse". As a result, the problem of instrumentalism arises, which means that it is the method as such that dictates how an Umwelt should be analyzed, so properties of the former indirectly shape properties of the latter [11,92]. For instance, application of a hierarchical classificatory algorithm necessarily provides a hierarchically arranged classification, even if the respective real diversity pattern can be non-hierarchical. Extreme formalizations implied by this idea are considered its main drawback from a biologically meaningful standpoint, as they presume inevitably undesirable reductions (see above).

One of the important methodological problems of the numerical program results from a variety of quantitative methods not reducible to either the most general or the most correct [25]. In such a pluralistic situation, the same question inevitably arises, as in the case of the logical taxonomic subprogram (see previous section). Now it is about selection of a particular method among several available. Two general solutions are possible here. On the one hand, the above principle of onto-epistemic correspondence can serve as a basis for such a selection. A method is preferable if it is more effective with respect to the biological content of the task being solved. On the other hand, the choice of a method should be justified "technologically". The better it is formulated within a well-founded axiomatics, the more preferable is it. The first approach is attractive from the point of view of biology, but it contradicts the ideology of the numerical program. The second approach, advocated by proponents of the "mathematical taxonomy" [93], as noted above, subordinates the solution of biologically sound tasks to the authority of the formal method and, thus, brings forth the problem of instrumentalism.

If philosophical questions are left aside, then undoubted practical advantages of the numerical program come to the fore. One of them is that numerical methods make possible comparative analyses of large data arrays. This is especially true for the numerical phyletics operating with many thousands of unit characters (nucleotide base pairs). A possibility of quantitative comparisons of different classifications by their characteristics, as well as elaboration of the consensus classifications for those derived from incompatible datasets, are also among the practical advantages of this program. At last, computer experimentations with virtual models make it possible to simulate macroscopic phenomena that are fundamentally unobservable and not amenable to direct experiments, such as the structure of biodiversity, global phylogeny, etc.

3.4. The Typological Program

The typological way of perceiving and representing the qualitative structure of the world is among the most basic aspects of cognitive activity [94]. It begins with personally perceiving and thinking of nature with gestalts, i.e., integrated images expressing essential features of its various manifestations (aspects, fragments, etc.). The results of such an intuitive perception are then transferred to nature itself. From this, an antique conception of archetype as an initial ideal form ("matrix"), giving rise to a diversity of all real forms, emerged. It also occurred in the general idea of the prototype underlying the natural philosophical concept of the Scala Naturae, which had a significant impact on the formation of the worldview among many naturalists of the 18th and 19th centuries [95].

Typological views are usually derived by authorities from the essentialist ones, but this is hardly true. Aristotle's understanding of essence (ousia), which forms the basis of essentialism in its widespread understanding (ascending basically to Popper), is functional [96,97]. It is this capacity that inspired many taxonomists from scholastics (such as Cesalpino and Tournefort) to early post-scholastics (such as Jussieu and Strickland). Contrary to this, the typology proper, as it appeared in the works of French and German anatomists at the end of the 18th and the beginning of the 19th centuries, was based on an idea of prototypes or archetypes understood structurally, as determined by spatial interrelations of the body parts of organisms [98,99]. Therefore, the typological program was undoubtedly an original product of the early post-scholastic development of systematics with one of its predecessors having been most likely an Aristotelian I. Jung with his conception of the geometric construction of plants (see Reference [11] for details). In the first half of the 19th century, this program dominated, especially in systematic zoology.

The central element of the typology in general, and of the typological theory in systematics in particular, is the type concept. The latter is very multifold and combines many different meanings reflected in a rather rich terminology, so such a kind of typological pluralism is to be taken into consideration when typology is discussed in general. Therefore, before exposing this program, it is appropriate to consider the main attributions of this concept.

In its broadest natural philosophical sense, a type is likened to a natural Law of Nature. According to the Swiss biologist A. Naef, "organisms relate to the type in the same way that events relate to the law they manifest" [100]. From this point of view, both a physical or a chemical law and a type thus understood are equally fundamental attributes of nature—though not directly observable and rather conceivable, but, nevertheless, completely material as metaphysical natural phenomena.

In a more concrete and yet quite natural philosophical understanding, a type is thought of as a kind of generalized structural characteristic of an organism, considered in a generalized (idealized) form. Such type can be expressed as a general body plan (Cuvier) or as a developmental plan (von Baer) or as a metamorphosis of parts of organismal archetype (Goethe). Thus understood, such a "natural philosophical (arche)type" plays a key role in the initial development of the concept of structural homology (R. Owen) without which no systematic (or any comparative) research is possible [33,34].

More empirically understood, a type corresponds to a combination of properties that are characteristic (typical) for a certain group of organisms (or eventually, any other objects). Such a group can be distinguished by a researcher on the basis of various reasons. It can be either a taxon in its proper typological understanding, or a monophyletic group, or a life form, or even just a phenon, etc. Thus, such an empirical type is largely viewpoint-dependent: it is the researcher who decides, guided by a particular concept, which kinds of groups are to be recognized and which properties are to be considered as constituting their types [101].

It is the natural philosophical concept of the archetype that is central to the typological research program in biological systematics. This program realizes a typological theory, according to which the Systema Naturae is a hierarchically ordered diversity of the (arche)types of various levels of generality, with the most fundamental taking the highest position in this hierarchy. This conception underlies a general idea of natural classification as the one most adequately representing the presumed hierarchical structure of the diversity of the archetypes. Respectively, taxa are recognized following the general principle of the unity of type. Each typological taxon is defined by an archetype of a certain level of generality most fully expressed in the organisms belonging to that taxon. For this, the characters used to recognize taxa are weighed and ranked in a special way: the most significant are those that allow us to characterize most completely the archetypes and their subordinations.

It is clear from the preceding that, in elaborating particular typological classifications, the analysis of archetypes is primary with respect to the classification of organisms [11,61,101]. This means that the hierarchy of archetypes is revealed first. Then the weighting/ranking of the characters attributed to them is carried out, and the typological taxa are diagnosed by these characters. In order to proceed properly from the hierarchy of archetypes through the analysis of the characters to the hierarchy of taxa, the principle of ranks coordination is introduced. Characters attributed to the archetypes of certain levels of generality are used to define the taxa of the same levels. This principle ascends to the methodologies of the French naturalists A.-L. de Jussieu and G. Cuvier. Like taxa of the onto-rational taxonomy (see Section 3.2), typological taxa are thought of quite realistically as the natural kinds. Accordingly, an objective status is also presumed for their taxonomic ranks [101].

The typological program most fully implements the general ideas of the natural philosophy-based typological theory (or theories) at the macro-systematic level, where differences in archetypes are most evident. At the lower (generic and especially species) levels, its capabilities are significantly lower, since the differences between organisms at these levels do not usually affect the body plans.

The natural philosophical typology has been developing from the very beginning in three main versions, which are known as stationary, epigenetic, and dynamic. They were added subsequently with several other versions, adapting the original ones to an evolutionary idea [11]. All of them might be treated as particular taxonomic theories of the typological kind.

For stationary typology, going back to the ideas of the French naturalists F. Vicq d'Azyr, G. Cuvier, and É. Geoffroy Saint-Hilaire, the central is an idea of a structural general plan (bauplan), determined by the spatial (geometric) relations of its components in adult organisms. The overall organismal diversity is structured by detailing these plans from the most to the least general. Accordingly, the typological unity of taxa appears as a unity of the structural plans of the organisms belonging to them.

In epigenetic typology (i.e., epitypology) originating from ideas of the German naturalist K. von Baer, the general plan is considered as coming through ontogenesis. The epigenetic type is mainly a type of individual development of an organismal structural plan. In modern terms, such a developmental type can be interpreted as an ontogenetic pattern. The general structure of the diversity of these patterns can be represented by a branching tree, which was suggested to call "Baerian" [102]. The general idea of the modern ontogenetic systematics is most close to this version of typology. In it, the diversity of taxa is analyzed from the standpoint of the diversity of the ontogenetic patterns of the respective organisms and the taxa are diagnosed by specificity of the patterns characteristic to their organisms [102–106].

Dynamic typology goes back to the ideas of the German poet and naturalist W. von Goethe. It considers the general organismic construction also in a development, but the latter is understood as an ideal (imaginary) metamorphosis (transformation) of different parts of an imaginary archetype of some superorganism. Therefore, it is sometimes called transformational typology [107]. In accordance with its principal idea, a taxon is characterized by a common pattern of particular transformations of the basic archetype. This typological theory served in the mid-19th century as a basis of the typological concept of homology (Owen). In the 20th century, it enjoyed popularity among some constructional morphologists [100,108–110].

From the second half of the 19th century, and especially in the 20th century, the typological program appeared to be in a "shadow" of the phylogenetic program and was almost completely rejected by the phenetic one. Two main arguments were put forward against it: (1) the type is an ideal construction, to which nothing corresponds in nature, and (2) the type is unchanged, which contradicts the central evolutionary idea of recent biology. To some extent, such a negative attitude is aggravated by the fact that the biochemical and especially molecular genetic attributes most preferred in contemporary systematic studies seem not to be amenable to the classical typological interpretation [111].

However, the traditionally negative attitude towards typology [4–6,8,112] began to be replaced gradually by a more positive one in the second half of the 20th century. Some of its key ideas were supported by the "new essentialism" [113,114], which is acknowledged to be compatible with the evolutionary ideas [115–117], up to a proposal of the phylogenetic typology concept [118]. Very interesting in this respect are evolutionary typological concepts of the dynamic archetype (phylocreod) [119–121] and the phylotype [122–124]. They refer to the stable trajectories of the evolutionary development of both particular morphological structures and integrated ontogenetic patterns. The most significant in its promising perspective seems to be a merging of the classical epitypological and phylogenetic ideas about historical formation of the structural plans of organisms with the most recent ideas on genetic regulation of ontogenesis within the framework of the evo-devo program in systematics (see Section 3.8).

3.5. The Biomorphic Program

In the basic structure of biodiversity, two principal components are usually recognized, namely, ecological and phylogenetic ones [125]. The former corresponds to the hierarchy of ecosystems, while the latter corresponds to the hierarchy of monophyletic groups. However, according to a wider concept, there are three such components. To those just indicated, the biomorphic component should be added, which is a hierarchy of biomorphs, or life forms [126]. The first component is studied by ecology and is outside of the scope of systematics. The second is explored by phylogenetics shaping a specific Umwelt for the phylogenetic program in systematics (see Section 3.7). The third component is explored by ecomorphology (in its taxonomic meaning, see below), or biomorphics. It is usually taken out of the limits of proper systematics, but recently was suggested to be included in the scope of the latter [11,127].

The need to develop natural classifications of the basic life forms rather than artificial diagnostic keys of scholastic systematics (like that of Linnaeus) was declared by the German naturalist A. von Humboldt at the beginning of the 19th century. This might have become one of the most important ideas in early post-scholastic systematics, but Humboldt's idea left no evident traces in the then prevalent taxonomic theories [11]. Sufficiently developed classifications of the life forms in botany and zoology began to appear in the late 19th and early 20th centuries [128,129]. Their main purpose was to reflect the diversity of the complexes of morphophysiological adaptations of living beings as elements of the ecosystems. By the end of the 20th century, a movement in this direction led to the emergence of a fairly developed theory, which was proposed to designate ecomorphology [130,131].

However, the latter term has two meanings. One of them is connected with an ecological interpretation of organismal morphology with its main task being an analysis of morphological adaptations [132]. Another is associated with the classification of organisms according to their ecomorphological (or biomorphological) similarities (references above). Taking this ambiguity into account, an introduction of the above term biomorphics to designate only the taxonomic aspect of ecomorphology and to consider it as one of the research programs within biological systematics seems justified.

The main task of the biomorphic program is to recognize the biomorphs of different levels of generality and to develop biomorphic classifications on this basis. Biomorphs are understood as groups of organisms distinguished by their bio-morphological (eco-morpho-physiological) specificity. Thus, biomorphically defined taxa unite organisms similar in their morpho-physiological features, which ensures fulfillment of similar ecosystem functions. Thus, the definition of a biomorph includes neither indication of kinship nor the time and place of the origin of organisms, nor their phenetic similarity as such.

Such interpretation of biomorphs provides the research program under consideration with a particular specificity that is not observed in other branches of biological systematics. As a matter of fact, different organisms of the same species, and even different stages of development of the same organism, can belong to different biomorphs if they differ significantly in their bio-morphological characteristics. For instance, biomorphologically different can be conspecific plants depending on their growth conditions, or insect larvae and images playing significantly different roles in the ecosystems. Thus, not the total organisms, but rather such elementary units ("bricks") of biomorphic diversity are united in the taxa in respective biomorphic classifications [133].

The elaboration of biomorphic classifications begins with character weighting and ranking. The most significant are those describing the most important adaptive features of organisms as elements of the ecosystems. Their ranks are determined by the levels of generality of the corresponding adaptations. In the analysis of characters, any distinction between homologies and analogies does not matter—the general morpho-physiological organization (an "adaptive syndrome") of living beings is considered instead. Based on the characters thus weighed and ranked, the entire classification of biomorphs is built up in which the hierarchy is determined by the hierarchy of respective morpho-physiological organizations of various levels of generality [130,131]. Thus, the classification algorithm of biomorphics is deductive: a common basis for dividing the world of living organisms (for example, a type of metabolism) is initially identified, and then the entire classification is construed from top to bottom following a hierarchy of the syndromes detailing the chosen basis successively.

Thus, the biomorphic program is very similar, by its general classificatory algorithm, to the typological program (see Section 3.4 on the latter). Their fundamental likeness is in considering a timeless aspect of diversity of organisms and classifying them based on a prior character weighing and ranking. A significant difference is that typological classifications are based mainly on homologies, while biomorphics takes into consideration the entire adaptive syndrome of features. At the same time, special emphasis on the functional significance of characters places biomorphics close to the Aristotelian essentialism (ousiology), in which particular attention is paid to the functional destiny of the organisms' parts (see Section 3.4).

From an ontological point of view, this research program takes a very realistic stance toward biomorphs. This is substantiated by reference with the processes in natural ecosystems shaping the entire structure of biomorphic diversity. With this, it is often assumed that there is only one single system of biomorphs because there is, supposedly, only one global functional structure of the entire biota [130,131]. This idea ascends evidently to the Humboldtian monistic natural philosophy. However, according to another point of view, it may make sense to develop different biomorphic classifications in which the same organisms can be allocated to different taxa [134–136]. All this seems to be similar, to a degree, to distinguishing between "general purpose" and "special purpose" classifications in the phenetic program (see Section 3.1) and means a certain balance between taxonomic monism and pluralism within biomorphics.

The program under consideration, by elaborating its biomorphic classifications, is of importance for that division of ecology dealing with analysis of the structure of ecosystems. The biomorphs recognized in such classifications relate, in a certain way, to classifications of the syntaxa considered, according to one of the synecological conceptions, as fundamental units of that structure [137]. On the other hand, properly construed biomorphic classifications provide very significant information for the analyses of the interrelation between structures of phylogenetic and biomorphic aspects of biodiversity, as they are shaped in the course of biological evolution.

3.6. The Biosystematic Program

The term biosystematics has two meanings. On the one hand, it is sometimes used to refer to the entirety of biological systematics: it is simply a contraction of these two words. On the other hand, this is one of the research programs in systematics dealing predominantly with the study of species and intraspecific diversity. In this paper, this term is used in the second sense.

Biosystematics thus understood appeared to be one of two main programs, along with phylogenetics (see the next section), implementing evolutionary ideas in biological systematics [138]. Its title biosystematics was meant to emphasize its main concern with the natural living populations, and not with the dead museum specimens, and it became likely the first to have been officially called evolutionary systematics [139]. Its principal conceptual source was the classificatory Darwinism of the second half of the 19th century, which was emphasized by calling it Darwinian systematics [140]. Accordingly, the beginners of this program declared persistently that the main lower taxonomic units are not "Linnaean species" but geographic races, which are the only natural biological entities deserving exploration and classification [138,140–144]. The term population systematics directly indicated the level in the taxonomic hierarchy it deals with [144]. At last, to stress a novelty of this program against the orthodox one, it was termed the "new systematics" [145–147].

Emergence of this taxonomic theory and program appeared to be, along with phenetics (see Section 3.1), a specific response of biological systematics to the challenges of the positivist philosophy of science. No less (if no more) important role in its shaping played an active formation of the new contents of biology at the beginning of the 20th century with its interest in ecology, physiology, genetics, and in its efforts to explain everything by evolutionary mechanisms acting at the population level. Biosystematics absorbed new ideas and facts, considering them from a standpoint of the evolution of populations. Due to this, it played an important role in the formation of an evolutionary concept called the "Modern Synthesis" in the 1930s–1940s [147].

According to the specific understanding of its tasks, biosystematics (together with phenetics) abandoned the general idea of the global natural system for the simple reason that supra-specific systematic categories were excluded from its particular Umwelt. It is mainly engaged in elucidating the ecological nature and genetic mechanisms of both the dynamics and the stability of intraspecific categories and units called gene-ecological by the Swedish biologist G. Turreson [148]. It was emphasized that these biosystematic units and their classifications should not necessarily coincide with those of "orthodox" systematics, since they were distinguished on different grounds [149,150]. It was also proposed, in addition to the classification of those units, to fix continuous trends ("clines") of geographic and ecotypic variability of particular characters over the entire ranges of widely distributed species [151,152].

Biosystematic studies focused on comparative analysis of data using all available categories of data to discriminate intraspecific gene-ecological units, thus realizing the phenetic idea. The only significant difference between phenetics and biosystematics, from a taxonomic perspective, is that the former uses only proper traits of organisms (see Section 3.1), while the latter pays attention to their ecological characteristics. This circumstance has predictably caused an extensive employment of numerical methods by biosystematics. In addition, within the framework of this program, a special area of research has been formed, namely, experimental systematics [153,154], which may be treated as a kind of response of systematics to the physicalist challenges, alongside with the above numerical program. It is based on an idea that all judgments about the differentiation of closely related species and intraspecific units should be subject to the experimental verifications under natural and/or laboratory conditions.

Biosystematic research, from the very beginning to the present, have been most popular in botany [154–160]. In particular, one of its leaders, the Soviet biologist A. Takhtadzhyan [157], defined it as "a branch of botany studying the taxonomic and population structure of species, its morphological, geographical, ecological, and genetic differentiation, origin, and evolution". In zoology, biosystematics (mostly under the name "new systematics") was initially promoted by E. Mayr [144], but, later, interest in it almost disappeared [161].

Recent phylogeography, dealing with reconstructions of the microphylogenies of widely distributed species [162–164], may have certain concern for the biosystematics issues. However, it restricts itself by the numerical phyletic methods and does not take into consideration other types of data (morphological, ecological, etc.). Therefore, its results play but an auxiliary role in complex biosystematic studies.

On the opposite side, the recently developed idea of the integrative systematics, as opposed to the total molecularization of research at the species level, can actually be considered a certain revival of the biosystematic theme in zoology. Its main idea is that the delimitation of species units by molecular genetic characters is only an initial stage in the study of species diversity in which the bulk should involve analysis of all available characters, which allows us to consider various aspects of species' natural history [165–168].

3.7. The Phylogenetic Program

This program is another version, along with biosystematics, of the implementation of evolutionary ideas in biological systematics, in this case at macro-evolutionary levels. The first attempt to initiate it was the "Philosophy of Zoology" by the French biologist Lamarck at the beginning of the 19th century. It was actually focused on macro-evolution but appeared to be premature. The second attempt was the "General Morphology of Organisms" by the German biologist E. Haeckel in the second half of the 19th century, which turned out to be much more successful. One of its principal outputs became systematic phylogeny (as Haeckel himself called it), associated with the historical interpretation of macro-taxa and their characters. It is now commonly known as phylogenetic systematics.

The main parameters of the Umwelt shaping the ontological basis of the phylogenetic theory and program can be briefly summarized and formalized as follows [26,27,169,170]. Phylogenetics is based on an assumption (quasi-axiom) that the ordered diversity of organisms is a result of the global long-term phylogenetic process encompassing biota as a whole. This process involves the origin of some groups (descendants) from others (ancestors), and the emergence of new groups being accompanied by the emergence of their specific properties (Darwin's formula descent with modification). It encompasses divergent (cladogenesis) and directed (anagenesis) components. Divergent evolution leads to an irreversible decrease in both kinship and similarity, while anagenetic evolution can lead to a secondary decrease in the similarity in some structures (convergence). Attributes of a newly emerging group of organisms are inherited from its closest common ancestor by the latter's descendants in both conserved and modified forms, and this makes them more similar to each other than to members of other groups (quasi-axiom of inherited similarity). Phylogeny thus understood produces a phylogenetic pattern defined as a hierarchy of monophyletic groups of different levels of generality interconnected by kinship (phylogenetic) relationships. It is evident that both the entire phylogenetic pattern and monophyletic groups within it are treated realistically.

From these basic assumptions, it is deduced that the natural classification should be phylogenetic. This means that any particular classification should reflect the structure of the respective phylogenetic pattern as completely as possible. This general idea is implemented by the principle of monophyly: a group should be characterized primarily by unity of origin, i.e., should include descendants of a single ancestral form. This principle is crucial for the entire phylogenetic program: only the monophyletic group (phylon), characterized by such a unity of origin, is thought to be natural and can be recognized as a taxon in phylogenetic classification. On the contrary, any group that does not meet this criterion is treated as polyphyletic and considered artificial in most schools of phylogenetic taxonomy. Accordingly, in elaborating a phylogenetic classification, the most significant characters are those that allow us to recognize monophyletic groups.

The main contribution of the phylogenetic program for the development of other branches of biology is that phylogenetic reconstructions provide a sufficiently reliable basis for the historical interpretations of similarities and differences between organisms by any kind of trait. One of the instruments of such an interpretation is the detection of the "phylogenetic signal" in the overall pattern of similarity relations, which means a measure of similarity of organisms due to their kinship relationships rather than to ecological causes [171]. Besides, phylogenetic reconstructions play a key role in the historical biogeography.

This program has been dominating in biological systematics since the mid-19th century. It was represented first by what can be reasonably termed classical (Haeckelian) phylogenetics. Two other main versions (subprograms) were added to it in the mid-20th century, namely, evolutionary taxonomy and cladistics. These subprograms basically differ in their treatments of the phylogenetic process (less or more reductionist), the relations between the phylogenetic pattern and the phylogenetic classification (less or more strict), as well as the methods of elaborating the latter (selection of characters, assessment of similarity, ranking taxa, etc.). Another important difference between them is determined by two particular interpretations of the principle of monophyly, which can be treated as either narrow or broad. In the first case (holophyly), a group is considered monophyletic if it includes all descendants of the same ancestor, with the latter being treated obligatory as a species. In the second case (paraphyly), a monophyletic group includes only part of the descendants of the same ancestor, which may be a supra-specific group. The groups defined according to these two versions of monophyly are termed holo- and paraphyletic, respectively. By all of these features, classical (Haeckelian) phylogenetics and evolutionary taxonomy are close to each other, while cladistics is the most specific.

The classical phylogenetic subprogram equally takes into account both cladogenetic and anagenetic components of phylogeny, though it does not place a particular emphasis on the adaptive nature of evolution. Evolutionary changes are considered mainly as transformations of the structural plans of organisms, according to which their groups are interpreted as various implementations of such plans. Thus, of great importance are the reconstructions of the ancestral body plans, which makes the Haeckelian approach a phylogenetic interpretation of structural typology [100] (see Section 3.4 on the latter). The geological time of existence of the groups is quite significant, as it allows us to treat some earlier organisms as potential ancestors of some more recent ones. The phylogenetic tree in its classical interpretation has a rather complicated configuration: it is "tied" to the geochronological scale and shows a sequence of divergence, time of existence, and dynamics of diversity of the monophyletic groups, as well as successive stages of transformations within respective prevailing anagenetic trends (such as "arthropodization", "mammalization", "angiospermization", etc.).

An emphasis on body plans implies that the monophyletic groups are characterized by commonality of both conservative and innovative characters, some of which can be acquired as a result of parallel evolution. Accordingly, monophyly is understood as "broad", so both holophyletic and paraphyletic groups are equally significant in the phylogenetic classifications of this kind. The main argument in favor of the reality of paraphyletic groups is that they do not lose their morphobiological specificity after cleavage of their "side branches" by developing their own novel specializations [172–177]. Examples are bryophytes and vascular plants, actinopterygians and tetrapods, archosaurian reptiles and birds, artiodactyls and cetaceans, etc.

The correspondence between the phylogenetic tree and a classification based on it is admitted to be soft. It is sufficient that the latter should be compatible with the branching structure of the tree and should not contain evident polyphyletic taxa (such as homoiotherms). Accordingly, the tree being converted into a classification can be cut in different fragments both vertically and horizontally to adequately reflect both the kinship relations and anagenetic specificity of the monophyletic groups. Therefore, generally speaking, the same phylogenetic tree can be equally represented by several phylogenetic classifications that have some different details. The hierarchy of phylogenetic classification in its classical interpretation is ranked.

The evolutionary taxonomy subprogram was designated by its founding father, the American biologist G. Simpson [178], in order to demarcate it terminologically from evolutionary systematics in its biosystematic interpretation (see previous section). This phylogenetic subprogram resembles the classical one by most of its key presumptions. Its specificity is determined by the great attention paid to the adaptive essence of the evolutionary process, with the concept of the adaptive zone playing an especially important role [179]. This makes evolutionary taxonomy similar, to an extent, to the above biomorphics. The most fundamental demarcation between them is defined by including quasi-axiom of evolution in shaping the former's Umwelt. The adaptive zone is defined as a set of environmental conditions that determine the general type of adaptation of organisms. With this, it is assumed that the morpho-physiological specificity of a group, acquired in the course of its evolution, is a result of a similar reaction of organisms with similar epigenetic organization, inherited from their common ancestor, to similar environmental conditions. According to this evolutionary model, the acquisition of a basic adaptative syndrome of a taxon due to parallel evolution of its members witnesses its evolutionary unity no less than the inherited features. Thus, taxonomic integrity is defined by three interrelated evolutionary parameters, which include the unity of origin (monophyly in its broad sense), the unity of morphobiological organization (anagenetic grade), and the unity of evolutionary trends (parallelisms) [178,180,181]. In arranging phylogenetic classifications, an auxiliary principle of decisive gap is introduced, according to which the levels of mutual distinctiveness of taxa should be taken into consideration in both their individuation and their ranking.

The cladistic subprogram, in contrast to the two just considered, is based on a drastically simplified representation of phylogeny, which is reduced to cladogenesis, and on a respectively simplified interpretation of both phylogenetic relations and phylogenetically significant similarity. The founding fathers of this version, the German biologists W. Zimmermann and W. Hennig, designated their approach as phylogenetic systematics [26,182–185] and this designation currently dominates [27,169,170,186,187]; sometimes it equates to the entirety of biological systematics [187–189]. However, as will be shown below, its differences from the other phylogenetic subprograms are so significant that the term cladistics proposed by the American biologist E. Mayr [190] is more than justified.

In cladistics, the phylogenetic tree is reduced to a fairly simple cladogram, and monophyly is refined to holophyly. The phylogenetic (more correctly, cladistic) unity was determined initially through a cladistic relationship as follows: two groups, A and B, are closer to each other than to another group, C, if the nearest common ancestor of A and B is more recent than the common ancestor of all three groups. In a later version currently dominating, this relationship is determined by reference not to a hypothetical ancestor, but to some real remote group. Two groups, A and B, constitute a holophyletic group if it is shown that they are sisters relative to a third group, C, external to them (routinely called an outgroup). Thus, the concept of ancestor, and, with it, the geological time scale are paradoxically excluded from an Umwelt of this branch of phylogenetics.

At an operational level, the principle of synapomorphy is introduced to reveal the hierarchy of sister groups, according to which the holophyletic group is determined only by synapomorphies, i.e., by similarities in apomorphic (uniquely derived) characters, while simplesiomorphy (similarity in ancestral and "parallel" characters) is not taken into account. This principle makes cladistic theory very peculiar with respect to its logistics [11,169,191]. In all other classificatory approaches, the two-state (Aristotelian) division logic dominates, in which judgments of "A" and "non-A" types (both presence and absence of characters) are equally significant for identifying taxa. In cladistics, the one-state logic of the Soviet logician Vasil'ev [192,193] actually operates. Only judgments of type "A" (synapomorphies) are significant, while judgments of type "non-A" (non-synapomorphies, i.e., simplesiomorphies) are insignificant for the recognition of holophyletic taxa. In addition, the above principle means that, for delineation of a holophyletic group, only its inner similarity is significant, while its outer differences from other taxa are insignificant. Thus, these two components of general similarity relations—similarity and difference—turn out to be logically asymmetric with respect to their classification function. For this reason, the above principle of the decisive gap is not relevant in cladistics. Lastly, the principle of synapomorphy replaces the typological component of classical phylogenetics with a variant of phenetic combinatorics of characters. A cladon (clade in a pure taxonomic sense) is identified more reliably if it is diagnosed with a larger number of synapomorphies ("the more characters, the better"), which presumes an active use of certain numerical methods.

Cladistic classification is based on a strict correspondence between the hierarchy of sister groups in a cladogram and the hierarchy of taxa in the respective classification. For this, the initial cladogram is cut vertically only, with all horizontal relations being discarded, which provides recognition of cladistically consistent taxa (cladons). This is complemented with the principle of equal ranking of sister groups: the groups descending to the same node (branching point) of a cladogram are treated in their respective classification as the taxa of the same rank. As a result, the ranked hierarchy of cladistic classifications for large diverse groups become very fractional and non-operational. This eventually leads to a suggestion to abandon fixed ranks from the hierarchy of cladistic classifications and to make them rankless [8,194–196].

The general position of cladistics regarding the ontological status of holophyletic groups is declared realistic [26,27,182–184]. However, as emphasized above (see Section 2.1), the more reduced the Umwelt constituting the ontological basis of a particular taxonomic theory is, the less the portion of objective reality of the original Umgebung it contains. This conclusion is clearly true in the case of cladistics: it is based on quite a reductionist representation of phylogeny, and, therefore, has a poorer ontology in comparison with both classical phylogenetics and evolutionary taxonomy.

In the contemporary phylogenetic studies of extant organisms, an approach called molecular phylogenetics (phylogenomics, genophyletics) takes a leading position. It includes the analysis of DNA or/and RNA nucleotide sequences, assessment of the similarity between organisms by these sequences, and construction of molecular phylogenetic trees based on this similarity. All these procedures employ numerical techniques, which makes numerical phyletics (see Section 3.3) an instrumental part of molecular phylogenetics. The transformation of molecular phylogenetic trees into classifications in practical studies strongly follows the above principles of cladistics. Thus, the molecular phylogenetics, from a taxonomic standpoint, can rightly be considered as part of the cladistic subprogram. Therefore, its taxonomic application is sometimes called genosystematics [11,197,198], even though it might be more correct to call it genocladistics.

It is curious enough that the most recent development of phylogenetic systematics means that the history of this biological discipline makes a kind of circle by returning to that stage when the scholastic genus–species scheme dominated [11,199]. One feature of this return is signified by an idea of rankless cladistic classifications, and another by using molecular genetic data exclusively for inferring these classifications, which revives something like a unified division basis.

According to the original intention of the cladism ideologists, their approach should be common for all living beings. This intention is implemented by the universal "Tree of Life" project [200]. However, as the recent results show, the hope for a universal "cladification" (see Reference [201] for this term) of the living matter is hardly warranted. The reason is that the basal fragment of the phylogenetic tree, shaped by the branching patterns of the prokaryote taxa, is not strictly divergent but, rather net-like [202]. This obstructs the elaboration of strictly "vertical" classifications that cladistics seek to achieve.

Currently, a conviction is gradually spreading among systematic theoreticians that cladistics, especially with its molecular appendage, is too reductionist to adequately reflect the structure of taxonomic diversity, even if the latter is simplified to being treated phylogenetically. This is reflected in the appearance of some publications speculating on possible developments of biological systematics beyond cladistics [199,203–205]. However, currently, the cladistic approach is actively developing at the methodological level and still dominating in practical systematic research.

3.8. In a Shade of Dominance: The Evo-Devo Program

Generally speaking, this research program is just beginning to take shape and is poorly noticeable against the currently dominating cladistics based on the analysis of molecular genetic data [199]. Its specificity is in that it focuses on the evolutionarily interpreted variety of ontogenetic patterns of multi-cellular organisms [206,207]. The basis for this is provided by a synthesis of the considered phylogenetics in its rather widened sense, epigenetic typology (see Section 3.4 on it), and the evo-devo concept (a well-known abbreviation for the evolutionary developmental biology). The concepts of dynamic archetype (phylocreod) and phylotype (phylotypic stage) mentioned above (see Section 3.4) fit well into the general context of this taxonomic theory. The first means a stable (canalized) trajectory of the evolutionary development of ontogenetic patterns, and the second refers to those patterns that are initial for particular phylocreods and changes of which lead to switching from one phylocrecode to another mainly due to changes in the composition and function of the regulatory genes.

From a historical perspective, this theory goes back to the classical principle of triple parallelism of the mid-19th century. It links (a) the distribution of the body plans of organisms in the natural system, (b) the sequence of appearance of organisms with various body plans in geochronology, and (c) the successions of ontogenetic stages in the individual development of those organisms. Its fundamental novelty is incorporation of the evo-devo concept that focuses on the historical changes of the genetic mechanisms of regulation of ontogenesis [208–213].

As can be seen, the evo-devo (or phylo-evo-devo) taxonomic theory and respective research program are based on a rather rich biologically meaningful ontology, which distinguishes it positively from reductionist cladistics and molecular phylogenetics. This means another, newer version (along with biomorphics and evolutionary taxonomy) of the most recent biologization of the systematics. At the same time, as far as phylogeny is considered one of the cornerstones of this program, it is possible to consider the latter as another branch of the phylogenetic program in its widest sense.

By focusing on the evolution of ontogenetic patterns and the epigenetic mechanisms ensuring their historical stability and dynamics, this research program brings its own version of representing historical patterns of biodiversity and respective classifications. The former can be represented by a phylo-ontogenetic tree, which is actually a phylogenetically interpreted "Baerian tree" (see Section 3.4 on the latter). This tree is transformed into a corresponding evo-devo classification in the same manner as the phylogenetic one with its ranking scale being derived from a sequence of appearances of respective ontogenetic patterns in the evolution of multicellular organisms. The main characteristics of an evo-devo taxon becomes its specific ontogenetic pattern as a whole dynamic system, not reducible to any particular developmental stages [103–105,207]. All this provides biological systematics with a rich ontological basis and allows it to get rid of the overload reductionism brought in it by the above "genocladistics". It deprives the molecular factology of its priority status and overcomes a kind of traditional adultocentrism in the consideration of organismal anatomy [212].

It is evident that the evo-devo research program is not universal: its application is limited to the groups of multicellular organisms with sufficiently developed ontogenetic cycles. Accordingly, many protists and apparently all prokaryotes appear to be outside the scope of its competence. However, this circumstance should hardly be considered a serious disadvantage. As emphasized above (see Section 2.2), any research program in systematics—more precisely, each particular taxonomic theory underlying it—is inevitably local with regard to its applications.

From an epistemological viewpoint, the research program under consideration faces a serious problem caused by its rich natural ontology. The latter presumes that the elaboration of evo-devo classifications should be based on a joint exploration of two complexly interacting multifaceted dynamic systems known as phylogeny and ontogeny [106,214,215]. In such a knotty cognitive situation, the so-called NP-completeness problem (see Babbitt [216]) becomes very relevant. This means that the more complicated the initial conditions of a certain research task are, the less likely it becomes to find its most optimal solution. In the case of systematics, this means a low probability to attain a classification most optimally representing a specific multi-faceted Umwelt shaped by the evolutionarily interpreted diversity of ontogenetic patterns [11,169]. Therefore, elaboration of the evo-devo natural classifications turns out to be significantly more troublesome as compared to phenetically or cladistically consistent ones. However, this problem is true for the evolutionary taxonomy (see previous section on it) as well, as it also deals with a very rich natural ontology.

At the moment, classifications realizing the evo-devo taxonomic theory most consistently and, thus, belonging to the program in question are very few [104,208–213,217,218]. The reason is that detailed studies on diversity and evolution of the mechanisms of regulation of ontogenesis in animals and plants on a modern epigenetic basis are just beginning. Therefore, as always occurs with new disciplines, they involve analyses of only a few model organisms. Therefore, it seems premature to consider how actively this research program will be developing, how productive it may turn out to be for systematics, how serious the alterations of taxonomic classifications may be, and which particular alterations will occur. Among the main tasks to be solved by the evo-devo taxonomic theory, to make the program in question more promising, seem to be the following: (1) elaboration of a calculus for assessment of the relative significance (weight) of the differences in molecular sequences and ontogenetic rearrangements, (2) elaboration of the general ranking scale for the evo-devo classifications of different groups of organisms, and (3) development of an optimal way to combine vertical and horizontal interrelations between groups with different ontogenetic patterns to reflect most adequately both their primitive (ancestral) and derived features.

4. Conclusions: How to Handle Taxonomic Pluralism

As emphasized at the very beginning of this article, taxonomic pluralism is supported by the conceptualism-based philosophy of science. Nevertheless, classical taxonomic monism is also very common, and not only in academic circles but also (perhaps even more so) among practitioners. Thus, the problem of their balance is very relevant for the contemporary systematics.

In a softer formulation, this problem can be represented not by confronting these positions but rather by a question as to how limited taxonomic pluralism could and should be [219,220]. Such a standpoint presumes implicitly that there are more or less good or bad taxonomic theories, so this issue actually refers to how to "separate the clean from the unclean" and to eliminate somehow the latter. This important question was considered briefly from a philosophical standpoint at the end of Section 2. In this section, it is the right place to consider it from a more empirical standpoint.

One of the main practical outputs of taxonomic pluralism at a theoretical level (multiplicity of taxonomic theories) is that it produces taxonomic pluralism at an empirical level (multiplicity of classifications for particular groups). The latter means that the same organisms can rightfully be allocated to different taxa in classifications based on different taxonomic theories. However, various users of taxonomic classifications wish to obtain a unified and stable list of taxa and do not intend to puzzle out differences between particular theories and research programs. Thus, practitioners seem to vote uniformly for taxonomic monism by supposing there is actually only one natural pattern of taxonomic diversity reflected by only one natural classification available for a uniform direct use in various applied projects.

The simplest and most straightforward answer is offered by a pragmatic perspective, according to which the main evaluating criterion for a taxonomic theory is the ability to convert it into an operational concept most effective in resolving certain practical tasks. Such a theory deserves development by providing support for the respective taxonomic community elaborating it, while others are doomed to elimination because of the restricted resources for systematic studies. Quite a demonstrative case in this respect is the recent short but hot discussion of the instability (plurality) of species classifications to be used for the conservation purposes [221–223]. It illustrates how such a "scientific social Darwinism" may turn the disagreement of scientific ideas into an administrative struggle for the limited resources, which, in most recent times, is promoted indirectly by the system of grant support for scientific activity [224].

Another approach, of an epistemological kind, appeals to the scientific consistency of a taxonomic theory that has to correspond to certain criteria, which allows it to distinguish scientific knowledge from others (religious, commonplace, etc.). In particular, it is presumed that such a theory should make it possible to elaborate testable scientific hypotheses about the structure of taxonomic diversity [219]. The problem in this case is that such criteria vary with evolution of the philosophy of science, so the theories consistent from one standpoint may appear to be inconsistent from another. Several decades ago, numerical phenetics pretended to be both the most scientifically consistent and the most effective, and struggled against phylogenetics [5]—and where is this theory now?

Lastly, it is possible to consider this question from a ontological (metaphysical) standpoint, according to which the best taxonomic theories are those that are substantiated by referencing the natural causes structuring biota, and, thus, are biologically meaningful enough (see Section 2.1). From such a perspective, preference should be given to the theories with well-developed natural ontologies modeling multifaceted taxonomic diversity as completely as possible. From this standpoint, any episto-rational theories (see Section 3.2 on these) are least relevant, whereas, among ontology-based theories, those developed by evolutionary taxonomy and evo-devo research programs seem to be more consistent and more effective as compared to, say, more reductionist cladistics (especially in its genosystematic version).

From a scientific perspective, any culling of some research programs in favor of others, by ascribing a privileged status to the latter, contradicts the fundamental principle of the freedom of scientific activity. From a metaphysical standpoint, the only serious limitation of a rampant pluralism of taxonomic theories seems to be imposed by the very most important task of biological systematics to produce biologically sound classifications representing most comprehensively multifaceted diversity of living matter. Theories that provide such a possibility are considered good while those that do not are considered bad. However, it should be borne in mind that such categorization of taxonomic theories should not be taken as universal; instead, they may appear to be either good or bad locally in different cognitive situations. This is because the structure of the diversity in different groups of organisms can be shaped by different causal factors, so the most pertinent (locally good) to them may be different partial taxonomic theories referring to different Umwelts. At any rate, the optimal theories seem to be those that (a) would embody the advances and diminish shortcomings of all three above ideas concerning the limitation of taxonomic pluralism, and (b) would be flexible enough to allow to take into account biological specifics of particular groups of organisms.

Facing irreducible multiplicity of the research programs in biological systematics, another more relevant question seems to come to the fore [11,17]. How should different classifications be combined, with each reflecting a particular manifestation of the entire taxonomic diversity, in order to get a whole picture of the latter? As indicated above (see Section 2.1), it hardly seems possible to elaborate a biologically sound integrated or "omnispective" classification. Thus, one of the possible answers to this question may be an appeal to develop something like a combined faceted classification. It would likely allow—for each group of organisms and, eventually, for the entire Tree of Life—to embody different particular classifications based on different taxonomic theories into a single pool. In this connection, one of the most pressing tasks of the general taxonomic theory would become the elaboration of an appropriate meta language with an exhaustive natural (non-formal) ontology uniting those developed under different particular theories.

Funding: The Governmental Theme no. AAAA-A16-116021660077-3.3 implemented by the Research Zoological Museum at Lomonosov Moscow State University, supported this contribution.

Acknowledgments: I am sincerely grateful to Alessandro Minelli for inviting me to contribute to the special issue on "Renegotiating Disciplinary Fields in the Life Sciences" and for commenting on a draft version of this contribution. Three peer reviewers are deeply thanked for their most useful criticisms, comments, and suggestions.

Conflicts of Interest: The author declares no conflict of interest.

References

1. Kellert, S.H.; Longino, H.E.; Waters, C.K. (Eds.) *Scientific Pluralism*; University of Minnesota Press: Minneapolis, MN, USA, 2006; ISBN 978-0-8166-4763-7.
2. Ruphy, S. *Scientific Pluralism Reconsidered: A New Approach to the (Dis)unity of Science*; University of Pittsburgh Press: Pittsburgh, PA, USA, 2016; ISBN 978-0-8229-4458-4.
3. Lakatos, I. *The Methodology of Scientific Research Programmes*; Cambridge University Press: New York, NY, USA, 1995; ISBN 978-0-5212-8031-0.
4. Ruse, M. *The Philosophy of Biology*; Hutchinson University Press: London, UK, 1973; ISBN 978-0-0911-5221-5.
5. Hull, D.L. *Science as a Process*; University Chicago Press: Chicago, IL, USA, 1988; ISBN 978-0-2263-6050-8.
6. Mayr, E. *Toward a New Philosophy of Biology*; Cambridge University Press: New York, NY, USA, 1988; ISBN 978-0-6748-9666-6.
7. Mahner, M.; Bunge, M. *Foundations of Biophilosophy*; Springer: Frankfurt, Germany, 1997; ISBN 978-3-6620-3368-5.
8. Ereshefsky, M. *The Poverty of the Linneaean Hierarchy: A Philosophical Study of Biological Taxonomy*; Cambridge University Press: New York, NY, USA, 2001; ISBN 978-0-5210-3883-6.

9. Richards, R.A. *The Species Problem: A Philosophical Analysis*; Cambridge University Press: New York, NY, USA, 2010; ISBN 978-0-521-19683-3.

10. Wilkins, J.S. *Species: A History of the Idea*; University California Press: Berkely, CA, USA, 2010; ISBN 978-0-520-26085-6.

11. Pavlinov, I.Y. *Foundations of Biological Systematics: History and Theory*; KMK Scientific Press: Moscow, Russia, 2018; ISBN 978-5-6040-7499-2. (In Russian, with English Content)

12. Mayr, E. Numerical phenetics and taxonomic theory. *Syst. Zool.* **1965**, *14*, 73–97. [CrossRef]

13. Hull, D.L. Contemporary systematic philosophies. *Ann. Rev. Ecol. Evol. Syst.* **1970**, *1*, 19–54. [CrossRef]

14. Pesenko, Y.A. Methodological analysis of systematics. I. Formulation of the problem, and principal taxonomic schools. In *Principles and Methods of Zoological Systematics*; Zoological Institute: Leningrad, Russia, 1989; ISBN 978-7925-0216-1. (In Russian, with English Summary)

15. Woodger, J.H. *The Axiomatic Method in Biology*; Cambridge University Press: Cambridge, UK, 1937.

16. Gregg, J.R. *The Language of Taxonomy*; Columbia University Press: New York, NY, USA, 1954.

17. Pavlinov, I.Y. How it is possible to elaborate taxonomic theory. *Zool. Issled.* **2011**, *10*, 45–100. (In Russian, with English Summary)

18. Swoyer, C. Conceptualism. In *Universals, Concepts, and Qualities: New Essays on the Meaning of Predicates*; Trawson, E.S., Chakrabarti, A., Eds.; CRC Press: Routledge, UK, 2006; pp. 127–154. ISBN 978-0-7546-5032-4.

19. Quine, W.V. *Ontological Relativity & Other Essays*; Columbia University Press: New York, NY, USA, 1996; ISBN 978-0231083577.

20. von Uexküll, J. The theory of meaning. In *Essential Readings in Biosemiotics. Anthology and Commentary*; Favareau, D., Ed.; Springer Science + Business Media: Dordrecht, The Netherlands, 2010; pp. 81–114. ISBN 978-1-4020-9649-5.

21. Kull, K. Umwelt and modelling. In *The Routledge Companion to Semiotics*; Cobley, P., Ed.; Routledge: London, UK, 2009; pp. 43–56.

22. Knyazeva, E.N. J. von Uexküll's Concept of Umwelt and its Significance for the Modern Epistemology. *Vopr. Filos.* **2015**, *5*, 30–44. (In Russian, with English Summary)

23. Wartofsky, M.W. *Models: Representation and Scientific Understanding*; Reidel: Dordrecht, The Netherlands, 1979.

24. Sokal, R.R.; Sneath, R.H.A. *Principles of Numercial Taxonomy*; W.H. Freeman & Co.: San Francisco, CA, USA, 1963.

25. Sneath, R.H.A.; Sokal, R.R. *Numercial Taxonomy. The Principles and Methods of Numerical Classification*; W.H. Freeman & Co.: San Francisco, CA, USA, 1973; ISBN 978-0-7167-0697-7.

26. Hennig, W. *Phylogenetic Systematics*; University Illinois Press: Urbana, IL, USA, 1966; ISBN 978-0-2520-6814-0.

27. Wiley, E.O. *Phylogenetics: The Theory and Practice of Phylogenetic Systematics*; John Wiley & Sons: New York, NY, USA, 1981; ISBN 978-0-4710-5975-2.

28. Voyshvillo, E.K. *Concept as a Form of Thinking: Logical and Epistemological Analysis*; Moscow State University Publ.: Moscow, Russia, 1989. (In Russian)

29. Hempel, G. *Aspects of Scientific Explanation and Other Essays in the Philosophy of Science*; Free Press: New York, NY, USA, 1965; ISBN 978-0-0291-4340-7.

30. Kosko, B. *Fuzzy Thinking: The New Science of Fuzzy Logic*; Hyperion: New York, NY, USA, 1993; ISBN 978-0-0065-4713-6.

31. Lyubarsky, G.Y. *Origins of Hierarchy: The History of Taxonomic Rank*; KMK Science Press: Moscow, Russia, 2018; ISBN 978-5-9500-8296-2. (In Russian)

32. de Beer, G. *Homology, an Unsolved Problem*; Oxford University Press: Oxford, UK, 1971.

33. Hall, B.K. (Ed.) *Homology: The Hierarchical Basis of Comparative Biology*; Academic Press: New York, NY, USA, 1994; ISBN 978-0-1231-8920-2.

34. Bock, G.R.; Cardew, G. (Eds.) *Homology*; John Wiley & Sons: Chichester, UK, 1999; ISBN 978-0-1231-9583-8.

35. Pavlinov, I.Y. The contemporary concepts of homology in biology: A theoretical review. *Biol. Bull. Rev.* **2012**, *2*, 36–54. [CrossRef]

36. Wilson, R.A. (Ed.) *Species: New Interdisciplinary Essays*; The MIT Press: Cambridge, MA, USA, 1999; ISBN 978-0-2627-3123-2.

37. Hey, J. *Genes, Categories, and Species. The Evolutionary and Cognitive Cause of the Species Problem*; Oxford University Press: New York, NY, USA, 2001; ISBN 978-0-1951-4477-2.

38. Stamos, D.N. *The Species Problem. Biological Species, Ontology, and the Metaphysics of Biology*; Lexington Books: Oxford, UK, 2003; ISBN 978-0-7391-0778-2.

39. Pavlinov, I.Y. (Ed.) *The Species Problem: Ongoing Issues*; InTech Open Access Publisher: Rijeka, Croatia, 2013; ISBN 978-953-51-0957-0.

40. Zachos, F.E. *Species Concepts in Biology. Historical Development, Theoretical Foundations and Practical Relevance*; Springer: Basel, Switzerland, 2016; ISBN 978-3-3194-4966-1.

41. Atran, S. *The Cognitive Foundations of Natural History: Towards an Anthropology of Science*; Cambridge University Press: New York, NY, USA, 1990; ISBN 978-0-5213-7293 0.

42. Berlin, B. *Ethnobiological Classification: Principles of Categorization of Plants and Animals in Traditional Societies*; Princeton University Press: Princeton, NJ, USA, 1992; ISBN 978-0-6910-9469-4.

43. Winsor, M.P. Setting up milestones: Sneath on Adanson and Mayr on Darwin. In *Milestones in Systematics*; Williams, D.M., Forey, P.L., Eds.; CRC Press: Boca Raton, FL, USA, 2019; pp. 1–18. ISBN 978-0-4152-8032-7.

44. Nelson, G. Cladistic analysis and synthesis: Principles and definitions, with a historical note on Adanson's Familles des Plantes (1763–1764). *Syst. Zool.* **1979**, *28*, 1–21. [CrossRef]

45. Gilmour, J.S.L. Taxonomy and philosophy. In *The New Systematics*; Huxley, J., Ed.; Oxford University Press: Oxford, UK, 1940; pp. 461–474.

46. Goodman, N. Seven strictures on similarity. In *Problems and Projects*; Bobs-Merrill: Indianapolis, IN, USA, 1972; pp. 437–446. ISBN 978-0-8144-1682-2.

47. Tversky, A. Features of similarity. *Psychol. Rev.* **1977**, *84*, 327–352. [CrossRef]

48. Sober, E. *Philosophy of Biology*, 2nd ed.; Westview Press: Boulder, CO, USA, 2000; ISBN 978-0-8133-4065-4.

49. Rieppel, O.; Kearney, M. Similarity. *Biol. J. Linn. Soc. Lond.* **2002**, *75*, 59–82. [CrossRef]

50. Kluge, A.G. Total evidence or taxonomic congruence: Cladistics or consensus classification. *Cladistics* **1998**, *14*, 151–158. [CrossRef]

51. Rieppel, O. The language of systematics, and the philosophy of "total evidence". *Syst. Biodivers.* **2004**, *2*, 9–19. [CrossRef]

52. Rieppel, O. The philosophy of total evidence and its relevance for phylogenetic inference. *Pap. Avulsos Zool. (Sao Paulo)* **2005**, *45*, 77–89. [CrossRef]

53. Popper, K. *Conjectures and Refutations: The Growth of Scientific Knowledge*; Routledge: London, UK, 2002; ISBN 978-0-7100-6507-0.

54. Newton-Smith, W.H. *The Rationality of Science*; Routledge & Kegan Paul Ltd.: Abingdon, UK, 1981; ISBN 0-203-04615-3.

55. Gaydenko, P.P. *Scientific Rationality and Philosophical Mind*; Progress-Traditsia: Moscow, Russia, 2003; ISBN 5-89826-142-7. (In Russian)

56. Pavlinov, I.Y. Concepts of rational taxonomy in biology. *Biol. Bull. Rev.* **2011**, *1*, 60–78. [CrossRef]

57. Lyubishchev, A.A. *Problems of the Form, System, and Evolution of Organisms*; Nauka: Moscow, Russia, 1982; p. 164. (In Russian)

58. Stevens, P.F.; Haüy, A.-P. Candolle: Crystallography, botanical systematics, and comparative morphology, 1780–1840. *J. Hist. Biol.* **1984**, *17*, 49–82. [CrossRef]

59. Drouin, J.-M. Principles and uses of taxonomy in the works of Augustin-Pyramus de Candolle. *Stud. Hist. Philos. Biol. Biomed. Sci.* **2001**, *32*, 255–275. [CrossRef]

60. Driesch, H. *The Science and Philosophy of the Organism*; Aberdeen University Print: Aberdeen, UK, 1908.

61. Meyen, S.V. The main aspects of typology of organisms. *Zh. Obshch. Biol.* **1978**, *39*, 495–508. (In Russian, with English Summary)

62. Quine, W.V. Natural kinds. In *Naming, Necessity and Natural Kinds*; Schwartz, S., Ed.; Cornell University Press: Ithaca, NY, USA, 1994; pp. 155–175. ISBN 978-0-8014-9861-9.

63. Linsky, B. Putnam on the meaning of natural kind terms. *Canad. J. Phil.* **1977**, *7*, 819–828. [CrossRef]

64. Popov, I.Y. "Periodical systems" in biology (a historical issue). *Verhandl. Gesch. Theor. Biol.* **2002**, *9*, 55–69.

65. Popov, I.Y. *Periodical Systems and a Periodical Law in Biology*; KMK Science Press: Moscow, Russia, 2008; ISBN 978-5-8731-7505-5. (In Russian, with English Summary)

66. Salthe, S. *Development and Evolution: Complexity and Change in Biology*; The MIT Press: Cambridge, MA, USA, 1993; ISBN 978-0-2621-9335-1.

67. Giampietro, M. Complexity and scales: The challenge for integrated assessment. *Integr. Assess.* **2002**, *3*, 247–265. [CrossRef]

68. Ho, M.W. How rational can rational morphology be? A post-Darwinian rational taxonomy based on a structuralism of process. *Theor. Biol. Forum* **1988**, *81*, 11–55.

69. Ho, M.W. An exercise in rational taxonomy. *J. Theor. Biol.* **1990**, *147*, 43–57. [CrossRef]

70. Ho, M.W.; Saunders, P.T. Rational taxonomy and the natural system – segmentation and phyllotaxis. In *Models in Phylogeny Reconstruction*; Scotland, R.W., Siebert, D.J., Williams, D.M., Eds.; Clarendon Press: Oxford, UK, 1994; pp. 113–124. ISBN 978-0-1985-4824-9.

71. Adanson, M. *Familles des Plantes*; Vincent: Paris, France, 1763.

72. Thompson, W.R. The philosophical foundations of systematics. *Canad. Entomol.* **1952**, *84*, 1–16. [CrossRef]

73. Pokrovsky, M.P. *Introduction to the Classiology*; Institute of Geology & Geochemistry: Ekaterinburg, Russia, 2014; ISBN 978-5-9433-2108-5. (In Russian)

74. Shuman, A.N. *Philosophical Logic: Origins and Evolution*; EkonomPress: Minsk, Belarus, 2001; ISBN 985-6479-26-6. (In Russian)

75. Russell, G. Logical pluralism. In *The Stanford Encyclopedia of Philosophy*, Summer 2019 ed.; Zalta, E.N., Ed.; 2019. Available online: https://plato.stanford.edu/archives/sum2019/entries/logical-pluralism/ (accessed on 10 January 2019).

76. Griffiths, G.C.D. On the foundations of biological systematics. *Acta Biotheor.* **1974**, *23*, 85–131. [CrossRef]

77. Gorham, G.; Hill, B.; Slowik, E.; Waters, C.K. *The Language of Nature: Reassessing the Mathematization of Natural Philosophy in the Seventeenth Century*; University of Minnesota Press: Minneapolis, MN, USA, 2016; ISBN 978-0-8166-9989-6.

78. Kant, I. The metaphysical faundatons of natural science. In *Philosophy of Science: An Historical Anthology*; McGrew, T., Alspector-Kelly, M., Allhoff, F., Eds.; Blackwell Publishing Ltd.: Oxford, UK, 2009; pp. 232–237. ISBN 978-1-4051-7542-5.

79. Barsanti, G. *La Scala, la Mappa, L'albero: Immagini e Classificazioni Della Natura Frasei e Ottocento*; Sansoni: Florence, Italy, 1992; ISBN 978-8-8383-1384-4.

80. Strickland, H.E. On the true method of discovering the natural system in zoology and botany. *Ann. Mag. Nat. Hist.* **1841**, *6*, 184–194. [CrossRef]

81. Smirnov, E.S. On construction of systematic categories. *Russ. Zool. J.* **1923**, *3*, 358–389. (In Russian)

82. Sterner, B. Well-structured biology: Numerical taxonomy's epistemic vision for systematics. In *Patterns in nature*; Hamilton, A., Ed.; University of California Press: Berkley, CA, USA, 2014; pp. 213–244. ISBN 978-0-7565-0246-1.

83. Quicke, D.L.J. *Principles and Techniques of Contemporary Taxonomy*; Chapman & Hall: London, UK, 1993; ISBN 978-9-4010-4945-0.

84. Abbot, L.A.; Bisby, F.A.; Rogers, D.J. *Taxonomic Analysis in Biology. Computers, Models, and Databases*; Columbia University Press: New York, NY, USA, 1985; ISBN 978-0-2310-4926-9.

85. Swofford, D.; Olsen, G.J.; Waddell, P.J.; Hillis, D.M. Phylogenetic inference. In *Molecular Systematics*, 2nd ed.; Hillis, D.M., Moritz, C., Mable, B.K., Eds.; Sinauer Association: Sunderland, MA, USA, 1996; pp. 407–514. ISBN 0-87893-282-8.

86. Nei, M.; Kumar, S. *Molecular Evolution and Phylogenetics*; Oxford University Press: Oxford, UK, 2000; ISBN 978-0-1951-3585-5.

87. Felsenstein, J. *Inferring Phylogenies*; Sinauer Association: Sunderland, MA, USA, 2004; ISBN 978-0-8789-3177-4.

88. Semple, C.; Steel, M. *Phylogenetics*; Oxford University Press: Oxford, UK, 2003; ISBN 978-0-1985-0942-4.

89. Shapiro, S. *Philosophy of Mathematics: Structure and Ontology*; Oxford University Press: New York, NY, USA, 1997; ISBN 978-0-1951-3930-3.

90. Perminov, V.Y. *Philosophy and Foundations of Mathematics*; Progress-Traditsia: Moscow, Russia, 2001; ISBN 5-89826-098-6. (In Russian)

91. Gillespie, A.; Cornish, F. Intersubjectivity: Towards a dialogical analysis. *J. Theor. Soc. Behav.* **2010**, *40*, 19–46. [CrossRef]

92. Rieppel, O. The nature of parsimony and instrumentalism in systematics. *J. Zool. Syst. Evol. Res.* **2007**, *45*, 177–183. [CrossRef]

93. Dunn, G.; Everitt, B.S. *An Introduction to Mathematical Taxonomy*; Cambridge University Press: New York, NY, USA, 1982; ISBN 978-0-4864-3587-9.

94. Hamlyn, D.W. *The Psychology of Perception. A Philosophical Examination of Gestalt Theory and Derivative Theories of Perception*; Routledge: London, UK, 2017; ISBN 978-1-1382-0275-7.

95. Diekmann, A. *Klassifikation, System, 'Scala Naturae'. Das Ordnen der Objekte in Naturwissenschaft und Pharmazie Zwischen 1700 und 1850*; Wissenschaftliche Verlag: Stuttgart, Germany, 1992; ISBN 978-3-8047-1213-3.

96. Charles, D. *Aristotle on Meaning and Essence*; Clarendon Press: Oxford, UK, 2000; ISBN 978-0-1992-5673-0.

97. Gotthelf, A. *Teleology, First Principles, and SCIENTIFIC method in Aristotle's Biology*; Oxford University Press: Oxford, UK, 2012; ISBN 978-0-1992-8795-6.

98. Shatalkin, A.I. Essentialism and typology. In *Contemporary Systematics: Methodological Aspects*; Pavlinov, I.Y., Ed.; Moscow University Press: Moscow, Russia, 1996; pp. 23–154. ISBN 978-5-8731-7617-5. (In Russian, with English Content)

99. Winsor, M.P. Non-essencialist methods in pre-Darwinian taxonomy. *Biol. Philos.* **2003**, *18*, 387–400. [CrossRef]

100. Naef, A. *Idealistische Morphologie und Phylogenetik (Zur Methodik der Systematischen Morphologie)*; Gustav Fischer: Jena, Germany, 1919.

101. Lyubarskii, G.Y. *Archetype, Style and Rank in Biological Systematics*; KMK Science Press: Moscow, Russia, 1996; ISBN 5-8731-7005-3. (In Russian)

102. Ho, M.W. Development, rational taxonomy and systematics. *Theor. Biol. Forum* **1992**, *85*, 193–211.

103. Orton, G.L. The role of ontogeny in systematics and evolution. *Evolution* **1955**, *9*, 75–83. [CrossRef]

104. Martynov, A.V. *Ontogenetic Systematics, and a New Model of Bilaterian Evoluiton*; KMK Science Press: Moscow, Russia, 2011; ISBN 978-5-8731-7750-9. (In Russian, with English Content)

105. Martynov, A.V. Ontogenetic systematics: The synthesis of taxonomy, phylogenetics, and evolutionary developmental biology. *Paleontol. J.* **2012**, *46*, 833–864. [CrossRef]

106. Pavlinov, I.Y. A critical analysis of A.V. Martynov's version of ontogenetic systematics. *Thalassas* **2013**, *29*, 23–33.

107. Zakharov, B.P. *Transformational Typological Systematics*; KMK Science Press: Moscow, Russia, 2005; ISBN 978-5-0411-5462-2. (In Russian)

108. Weber, H. Konstrtionsmorphologie. *Zool. Jahrb. Abt. Anat. Ontog. Tiere* **1958**, *68*, 1–112.

109. Schmidt-Kittler, N.; Vogel, K. (Eds.) *Constructional Morphology and Evolution*; Springer: Heidelberg, Germany, 1991; ISBN 978-3-6427-6156-0.

110. Beklemished, V.N. *Methodology of Systematics*; KMK Scientific Press: Moscow, Russia, 1994; ISBN 5-8731-7005-3. (In Russian)

111. Shatalkin, A.I. Problem of archetype and contemporary biology. *Zh. Obshch. Biol.* **2002**, *63*, 275–291. (In Russian, with English Summary) [PubMed]

112. Hull, D.L. The effect of essentialism on taxonomy: Two thousand years of stasis. *Br. J. Philos. Sci.* **1965**, *15*, 314–326. [CrossRef]

113. Ellis, B. *Scientific Essentialism*; Cambridge University Press: Cambridge, UK, 2001; ISBN 978-0-5218-0094-5.

114. Rieppel, O. New essentialism in biology. *Philos. Sci.* **2010**, *77*, 662–673. [CrossRef]

115. Walsh, D. Evolutionary essentialism. *Br. J. Philos. Sci* **2006**, *57*, 425–448. [CrossRef]

116. Lewens, T. Evo-devo and "typological thinking": An exculpation. *J. Exp. Zool.* **2009**, *312B*, 789–796. [CrossRef]

117. Riegner, M.F. Ancestor of the new archetypal biology: Goethe's dynamic typology as a model for contemporary evolutionary developmental biology. *Stud. Hist. Philos. Biol. Biomed. Sci.* **2013**, *44*, 735–744. [CrossRef]

118. Vasil'eva, L.N. Hierarchical model of evolution. *Zh. Obshch. Biol.* **1998**, *59*, 5–23. (In Russian, with English Summary)

119. Waddington, C.H. *New Pattern in Genetics and Development*; Columbia University Press: New York, NY, USA, 1962; ISBN 978-1-1252-7834-5.

120. Waddington, C.H. *Towards a Theoretical Biology: Prolegomena*; Edinburgh University Press: Edinburgh, Scotland, 1968; ISBN 978-0-8522-4018-2.

121. Meyen, S.V. Plant morphology in its nomothetical aspects. *Bot. Rev.* **1973**, *39*, 205–260. [CrossRef]

122. Slack, J.M.W.; Holland, P.W.H.; Graham, C.F. The zootype and the phylotypic stage. *Nature* **1993**, *361*, 490–492. [CrossRef] [PubMed]

123. Hall, B.K. Baupläne, phylotypic stages, and constraint: Why there are so few types of animals. *Evol. Biol.* **1996**, *29*, 251–261.

124. Richardson, M.K.; Minelli, A.; Coates, M.; Hanken, J. Phylotypic stage theory. *Trends Ecol. Evol.* **1998**, *13*, 158. [CrossRef]

125. Faith, D.P. Biodiversity. In *The Stanford Encyclopedia of Philosophy*, Summer 2003 ed.; Zalta, E.N., Ed.; 2003. Available online: http://plato.stanford.edu/archives/sum2003/entries/biodiversity/ (accessed on 4 December 2007).

126. Pavlinov, I.Y. On the structure of biodiversity: Some metaphysical essays. In *Focus on Biodiversity Research*; Schwartz, J., Ed.; Nova Science Publishers: New York, NY, USA, 2007; pp. 101–114. ISBN 978-1-6002-1372-4.

127. Pavlinov, I.Y. Comments on biomorphics (ecomorphological systematics). *Zh. Obshch. Biol.* **2010**, *71*, 187–192. (In Russian, with English Summary)

128. Warming, E. *Om Planterigest Lifsformer*; Festskr. udg. University Kjobenhavn: Kjobenhavn, Denmark, 1908.

129. Friederichs, F.C. *Die Grundfragen und Gesetzmässigkeiten der Land- und Forstwirtschaftlichen Zoologie, Insbesondere der Entomologie*; Parey: Berlin, Germany, 1930.

130. Ale'ev, Y.G. *Ecomorphology*; Naukova Dumka: Kiev, Ukraine, 1986. (In Russian)

131. Leont'ev, D.V.; Akulov, A.Y. Ecomorphema of the organic world: An experience of construing. *Zh. Obshch. Biol.* **2004**, *65*, 500–526. (In Russian, with English Summary)

132. Bock, W.J. Concepts and methods in ecomorphology. *J. Biosci.* **1994**, *19*, 403–413. [CrossRef]

133. Kirpotin, S.N. Life forms of organisms as patterns of organization and spatial environmental factors. *Zh. Obshch. Biol.* **2005**, *66*, 239–250. (In Russian, with English Summary)

134. Du Rietz, G.E. Life forms of terrestrial flowering plants. *Acta Phytogeogr. Suec.* **1931**, *3*, 1–95.

135. Remane, A. Die Bedeutung der Lebensformtypen für die Orologi. *Biol. Gen.* **1943**, *17*, 164–182.

136. Serebryakov, I.G. *Ecological Morphology of Plants. Life Forms of Angiosperms and Conifers*; Vys'shaya Shkola: Moscow, Russia, 1962. (In Russian)

137. Mirkin, B.M. *Theoretical Foundations of Contemporary Phytocenology*; Nauka: Moscow, Russia, 1985. (In Russian)

138. Camp, W.H. Biosystematy. *Brittonia* **1951**, *7*, 113–127. [CrossRef]

139. Hall, H.M.; Clements, F.E. The phylogenetic method in taxonomy: The North American species of Artemisia, Chrysothamnus, and Atriplex. *Publ. Carnegie Inst. Wash.* **1923**, *326*, 5–355.

140. Komarov, V.L. Species and its subdivisions. *Dnevnik 11 S'ezda Russ. Estestvoisp. Vrach.* **1902**, *6*, 250–252.

141. Turrill, W.B. Species. *J. Bot. Lond.* **1925**, *63*, 359–366.

142. Huxley, J. (Ed.) *The New Systematics*; Oxford University Press: London, UK, 1940.

143. Mayr, E. *Systematics and the Origin of Species, from the Viewpoint of Zoologist*; Columbia University Press: New York, NY, USA, 1942.

144. Mayr, E. *Principles of Systematic Zoology*; McGrow Hill Book Co.: New York, NY, USA, 1969; ISBN 978-0-0704-1143-2.

145. Hubbs, C.L. Racial and individual variation in animals, especially fishes. *Am. Nat.* **1934**, *68*, 115–128. [CrossRef]

146. Turrill, W.B. The expansion of taxonomy with special reference to spermatophyta. *Biol. Rev.* **1938**, *13*, 342–373. [CrossRef]

147. Huxley, J.S. *Evolution: The Modern Synthesis*; G. Allen & Unwin Ltd.: London, UK, 1942.

148. Turreson, G. The species and the varieties as ecological units. *Hereditas* **1922**, *3*, 100–113. [CrossRef]

149. Sylvester-Bradley, P.C. *The Classification and Coordination of Infraspecific Categories*; Systematic Association: London, UK, 1952.

150. Valentine, D.H.; Löve, A. Taxonomic and biosystematic categories. *Brittonia* **1958**, *10*, 153–166. [CrossRef]

151. Huxley, J.S. Clines: An auxiliary method in taxonomy. *Bijdr. Dierk.* **1939**, *27*, 491–520. [CrossRef]

152. Endler, J.A. *Geographic Variation, Speciation and Clines*; Princeton University Press: Princeton, NJ, USA, 1977; ISBN 978-0-6910-8192-2.

153. Heslop-Harrison, J. *New Concepts in Flowering-Plant Taxonomy*; Harvard University Press: Cambridge, MA, USA, 1960; ISBN 978-0-4356-1390-7.

154. Hoch, P.C.; Stephenson, A.G. (Eds.) *Experimental and Molecular Approaches to Plant Biosystematics*; Missouri Botanical Garden: St. Louis, IL, USA, 1995; ISBN 978-0-9152-7930-2.

155. Davis, P.H.; Heywood, V.H. *Principles of Angiosperm Taxonomy*; Oliver & Boyd: London, UK, 1963; ISBN 978-8-1701-9383-8.

156. Solbrig, O.T. *Principles and Methods of Plant Biosystematics*; Macmillan Co.: New York, NY, USA, 1970; ISBN 978-0-0241-3700-5.

157. Takhtajan, A.L. Biosystematics: Past, present and future. *Bot. Zh.* **1970**, *55*, 331–345. (In Russian)

158. Lines, J.L.; Mertens, T.R. *Principles of Biosystematics*; Educational Methods: Chicago, IL, USA, 1970; ISBN 978-1-6828-6265-0.

159. Stace, C.A. *Plant. Taxonomy and Biosystematics*, 2nd ed.; Cambridge University Press: London, UK, 1989; ISBN 978-0-5214-2785-2.

160. Stuessy, T.F. *Plant. Taxonomy. The Systematic Evaluation of Comparative Data*, 2nd ed.; Columbia University Press: New York, NY, USA, 2008; ISBN 978-0-2311-4712-5.

161. Mayr, E.; Ashlock, P. *Principles of Systematic Zoology*, 2nd ed.; McGrow Hill Book Co.: New York, NY, USA, 1991; ISBN 978-0-0704-1144-9.

162. Avise, J.C. *Phylogeography. The History and Formation of Species*; Harvard University Press: Cambridge, MA, USA, 2000; ISBN 978-0-6746-6638-2.

163. Avise, J.C. Phylogeography: Retrospect and prospect. *J. Biogeogr.* **2009**, *36*, 3–15. [CrossRef]

164. Gutiérrez-García, T.A.; Vázquez-Domínguez, E. Comparative phylogeography: Designing studies while surviving the process. *BioScience* **2011**, *61*, 857–868. [CrossRef]

165. Dayrat, B. Towards integrative taxonomy. *Biol. J. Linn. Soc. Lond.* **2005**, *85*, 407–415. [CrossRef]

166. Padial, J.M.; Miralles, A.; Riva, I.; Vences, M. The integrative future of taxonomy. *Front. Zool.* **2010**, *7*, 16. [CrossRef] [PubMed]

167. Schlick-Steiner, B.C.; Steiner, F.M.; Seifert, B.; Stauffer, C.; Erhard, C.; Ross, H.C. Integrative taxonomy: A multisource approach to exploring biodiversity. *Ann. Rev. Entomol.* **2010**, *55*, 421–438. [CrossRef] [PubMed]

168. Goulding, T.C.; Dayrat, B. Integrative taxonomy: Ten years of practice and looking into the future. In *Aspects of Biodiversity*; Pavlinov, I.Y., Kalyakin, M.V., Sysoev, A.V., Eds.; KMK Science Press: Moscow, Russia, 2016; pp. 116–133. ISBN 978-5-9908-4166-6.

169. Pavlinov, I.Y. *Introduction to Contemporary Phylogenetics*; KMK Scientific Press: Moscow, Russia, 2005; ISBN 978-5-0411-5481-3. (In Russian, with English Summary)

170. Wägele, J.-W. *Foundations of Phylogenetic Systematics*; Friedrich Pfeil Verlag: Munchen, Germany, 2005; ISBN 978-3-8993-7056-0.

171. Revell, L.J. Phylogenetic signal, evolutionary process, and rate. *Syst. Biol.* **2008**, *57*, 591–601. [CrossRef] [PubMed]

172. Tatarinov, L.P. Classification and phylogeny. *Zh. Obshch. Biol.* **1977**, *38*, 676–689. (In Russian, with English Summary)

173. Saether, O.A. Underlying synapomorphies and anagenetic analysis. *Zool. Script.* **1979**, *8*, 305–312. [CrossRef]

174. Rasnitsyn, A.P. Conceptual issues in phylogeny, taxonomy, and nomenclature. *Contribut. Zool.* **1996**, *66*, 3–41. [CrossRef]

175. Rieppel, O. Monophyly, paraphyly, and natural kinds. *Biol. Philos.* **2005**, *20*, 465–487. [CrossRef]

176. Hörandl, E. Paraphyletic versus monophyletic taxa—Evolutionary versus cladistic classification. *Taxon* **2006**, *55*, 564–570. [CrossRef]

177. Stuessy, T.F.; Hörandl, E. Evolutionary systematics and paraphyly: Introduction. *Ann. Missouri Bot. Gard.* **2014**, *100*, 2–5. [CrossRef]

178. Simpson, G.G. *Principles of Animal Taxonomy*; Columbia University Press: New York, NY, USA, 1961; ISBN 978-0-2310-2427-3.

179. Van Valen, L.M. Adaptive zones and the orders of mammals. *Evolution* **1971**, *25*, 420–428. [CrossRef] [PubMed]

180. Bock, W. Philosophical foundations of classical evolutionary classification. *Syst. Zool.* **1974**, *11*, 375–392. [CrossRef]

181. Bock, W.J. Foundations and methods of evolutionary classification. In *Major Patterns of Vertebrate Evolution*; Hecht, M.K., Goody, P.C., Hecht, B.M., Eds.; Plenum Press: New York, NY, USA, 1977; pp. 851–895. ISBN 978-0-3063-5614-8.

182. Zimmermann, W. Arbeitsweise der botanischen Phylogenetik und anderer Gruppierungswissenschaften. In *Handbuch der Biologischen Arbeitsmethoden*; Abderhalden, E., Ed.; Urban & Schwarzenberg: Berlin, Germany, 1931.

183. Zimmermann, W. Die Methoden der Phylogenetik. In *Die Evolution der Organismen*; Heberer, G., Ed.; G. Fischer: Jena, Germany, 1943; pp. 20–56.

184. Hennig, W. *Grundzuge Einiger Theorie der Phylogenetische Systematik*; Deutscher Zentralverlag: Berlin, Germany, 1950.

185. Hamilton, A. (Ed.) *The Evolution of Phylogenetic Systematics*; University of California Press: Berkley, CA, USA, 2014; ISBN 978-0-5202-7658-1.

186. Rieppel, O. *Phylogenetic Systematics. Haeckel to Hennig*; Taylor & Francis: Boca Raton, FL, USA, 2016; ISBN 978-0-3678-7645-6.

187. Shatalkin, A.I. *Biological Systematics*; Moscow University Press: Moscow, Russia, 1988; ISBN 978-5-2110-0145-9.

188. Schuh, R.T. *Biological Systematics. Principles and Applications*; Cornell University Press: Ithaca, NY, USA, 2000; ISBN 978-0-8014-4799-0.

189. Williams, D.M.; Ebach, M.C. *Foundations of Systematics and Biogeography*; Springer Science + Business Media: New York, NY, USA, 2008; ISBN 978-1-4419-4445-0.

190. Mayr, E. Cladistic analysis or cladistic classification? *Zeitschr. Zool. Syst. Evol. Forsch.* **1974**, *12*, 94–128. [CrossRef]

191. Pavlinov, I.Y. *Cladistic Analysis (Methodological Problems)*; Moscow University Press: Moscow, Russia, 1990; ISBN 5-211-00918-5. (In Russian)

192. Vasil'ev, N.A. *Imaginary Logic*; Nauka: Moscow, Russia, 1989. (In Russian)

193. Bazhanov, V.A.N.A. *Vasil'ev and His Imaginary Logic. Resurrection of One Forgotten Idea*; Canon+: Moscow, Russia, 2009; ISBN 978-5-8837-3196-8. (In Russian, with English Summary)

194. Løvtrup, S. Phylogenetics: Some comments on cladistic theory and methods. In *Major Patterns of Vertebrate Evolution*; Hecht, M.K., Goody, P.C., Hecht, B.M., Eds.; Plenum Press: New York, NY, USA, 1977; pp. 805–822. ISBN 978-0-3063-5614-8.

195. de Queiroz, K.; Gauthier, J. Phylogenetic taxonomy. *Ann. Rev. Ecol. Evol. Syst.* **1992**, *23*, 449–480. [CrossRef]

196. Mishler, B.D.; Wilkins, J.S. The hunting of the SNaRC: A snarky solution to the species problem. *Philos. Theor. Pract. Biol.* **2018**, *10*, 1–18. [CrossRef]

197. Ratner, V.A.; Zharkikh, A.A.; Kolchanov, N.; Rodin, S.N.; Solovyov, V.V.; Antonov, A.S. *Molecular Evolution*; Springer: Berlin & Heidelberg, Germany, 1996; ISBN 978-3-5405-7083-7.

198. Antonov, A.S. *Plant. Genosystematics*; Akademkniga: Moscow, Russia, 2006; ISBN 5-9462-8271-9. (In Russian, with English Summary)

199. Pavlinov, I.Y. *Biological Systematics: In Search of the Natural System*; KMK Scientific Press: Moscow, Russia, 2019; ISBN 978-5-907099-95-1. (In Russian)

200. Cracraft, J.; Donoghue, M.J. (Eds.) *Assembling Tree of Life*; Oxford University Press: Oxford, UK, 2004; ISBN 978-0-1951-7234-8.

201. Mayr, E.; Bock, W.J. Classifications and other ordering systems. *J. Zool. Syst. Evol. Res.* **2002**, *40*, 169–194. [CrossRef]

202. Bapteste, E.; O'Malley, M.A.; Beiko, R.G.; Ereshefsky, M.; Gogarten, P.; Franklin-Hall, L.; Lapointe, F.-J.; Dupré, J.; Dagan, T.; Boucher, Y. Prokaryotic evolution and the tree of life are two different things. *Biol. Direct.* **2009**, *4*, 34. [CrossRef]

203. Wheeler, Q.D. (Ed.) *The New Taxonomy*; CRC Press: Boca Raton, FL, USA, 2008; ISBN 978-0-8493-9088-3.

204. Williams, D.M.; Knapp, S. (Eds.) *Beyond Cladistics: The Branching of a Paradigm*; University California Press: Berkeley, CA, USA, 2010; ISBN 978-0-5202-6772-5.

205. Hörandl, E. Beyond cladistics: Extending evolutionary classifications into deeper time levels. *Taxon* **2010**, *59*, 345–350. [CrossRef]

206. Minelli, A. Phylo-evo-devo: Combining phylogenetics with evolutionary developmental biology. *BMC Biol.* **2009**, *7*, 36. [CrossRef]

207. Minelli, A. Biological systematics in the evo-devo era. *Europ. J. Taxon.* **2015**, *125*, 1–23. [CrossRef]

208. Minelli, A. *The Development of Animal Form: Ontogeny, Morphology, and Evolution*; Cambridge University Press: Cambridge, MA, USA, 2003; ISBN 978-0-5110-7241-3.

209. Carroll, S.B. *Endless Forms most Beautiful: The New Science of Evo Devo and the Making of the Animal Kingdom*; Weidenfeld & Nicolson: London, UK, 2005; ISBN 978-0-297-85094-6.

210. Laubichler, M.D.; Maienschein, J. (Eds.) *Evolving Pathways: Key Themes in Evolutionary Developmental Biology*; Cambridge University Press: New York, NY, USA, 2008; ISBN 978-0-521-88024-4.

211. Minelli, A.; Pradeu, T. (Eds.) *Towards a Theory of Development*; Oxford University Press: London, UK, 2014; ISBN 978-0-1996-7143-4.

212. Moczek, A.P.; Sears, K.E.; Stollewerk, A.; Wittkopp, P.J.; Diggle, P.; Dworkin, I.; Ledon-Rettig, C.; Matus, D.Q.; Roth, S.; Abouheif, E.; et al. The significance and scope of evolutionary developmental biology: A vision for the 21st century. *Evol. Dev.* **2015**, *17*, 198–219. [CrossRef] [PubMed]

213. Rieppel, O. Ontogeny, phylogeny, and classification. In *Phylogeny and the Classification of Fossil and Recent Organisms*; Schmidt-Kittler, N., Willmann, R., Eds.; Verlag Paul Parey: Hamburg, Germany, 1989; pp. 63–82. ISBN 978-3-4901-4496-6.

214. Rieppel, O. Ontogeny—A way forward for systematics, a way backward for phylogeny. *Biol. J. Linn. Soc. Lond.* **2008**, *39*, 177 191. [CrossRef]

215. Garey, M.R.; Johnson, D.S. *Computer and Intractability: A Guide to the Theory of NP-Completeness*; W.H. Freeman & Co.: San Francisco, CA, USA, 1979; ISBN 978-0-7167-1045-5.

216. Babbitt, C.C. Developmental Systematics: Synthesizing Ontogeny and Phylogeny in the Malacostraca (Crustacea). Ph.D. Thesis, University of Chicago, Chicago, IL, USA, 2005. Available online: https://www.researchgate.net/publication/35700543_Developmental_systematics_synthesizing_ontogeny_and_phylogeny_in_the_malacostraca_crustacea (accessed on 1 June 2005).

217. Martynov, A.; Ishida, Y.; Irimura, S.; Tajiri, R.; O'Hara, T.; Fujita, T. When ontogeny matters: A new Japanese species of brittle star illustrates the importance of considering both adult and juvenile characters in taxonomic practice. *PLoS ONE* **2015**, *10*, e0139463. [CrossRef] [PubMed]

218. Ereshefsky, M. Eliminative pluralism. *Philos. Sci.* **1992**, *59*, 671–690. [CrossRef]

219. Minelli, A. Taxonomy needs pluralism, but a controlled and manageable one. *Megataxa* **2020**, *1*, 9–18. [CrossRef]

220. Garnett, S.T.; Christidis, L. Taxonomy anarchy hampers conservation. *Nature* **2017**, *546*, 25–27. [CrossRef]

221. Raposo, M.A.; Stopiglia, R.; Brito, G.R.R.; Bockmann, F.A.; Kirwan, G.M.; Gayon, J.; Dubois, A. What really hampers taxonomy and conservation? A riposte to Garnettand Christidis (2017). *Zootaxa* **2017**, *4317*, 179–184. [CrossRef]

222. Thomson, S.A.; Pyle, R.L.; Ahyong, S.T.; Alonso-Zarazaga, M.; Ammirati, J.; Araya, J.F.; Ascher, J.S.; Audisio, T.L.; Azevedo-Santos, V.M.; Bailly, N.; et al. Taxonomy based on science is necessary for global conservation. *PLoS Biol.* **2018**, *16*, e2005075. [CrossRef]

223. Khaitun, S.D. *The Crisis of Science as a Mirror Reflection of the Crisis of the Theory of Knowledge*; LENAND: Moscw, Russia, 2014; ISBN 978-5-9710-2665-5.

224. Broughton, V. The need for a faceted classification as the basis of all methods of information retrieval. *New Inform. Persp.* **2006**, *58*, 49–72. [CrossRef]

 *philosophies*

Article

EvoDevo: An Ongoing Revolution?

Salvatore Ivan Amato

Department of Cognitive Sciences, University of Messina, Via Concezione 8, 98121 Messina, Italy;
samato@unime.it or ivan.amato7@gmail.com

Received: 27 September 2020; Accepted: 2 November 2020; Published: 5 November 2020

Abstract: Since its appearance, Evolutionary Developmental Biology (EvoDevo) has been called an emerging research program, a new paradigm, a new interdisciplinary field, or even a revolution. Behind these formulas, there is the awareness that something is changing in biology. EvoDevo is characterized by a variety of accounts and by an expanding theoretical framework. From an epistemological point of view, what is the relationship between EvoDevo and previous biological tradition? Is EvoDevo the carrier of a new message about how to conceive evolution and development? Furthermore, is it necessary to rethink the way we look at both of these processes? EvoDevo represents the attempt to synthesize two logics, that of evolution and that of development, and the way we conceive one affects the other. This synthesis is far from being fulfilled, but an adequate theory of development may represent a further step towards this achievement. In this article, an epistemological analysis of EvoDevo is presented, with particular attention paid to the relations to the Extended Evolutionary Synthesis (EES) and the Standard Evolutionary Synthesis (SET).

Keywords: evolutionary developmental biology; evolutionary extended synthesis; theory of development

1. Introduction: The House That Charles Built

Evolutionary biology, as well as science in general, is characterized by a continuous genealogy of problems, i.e., "a kind of dialectical sequence" of problems that are "linked together in a continuous family tree" [1] (p. 148).

Let's start with Charles Darwin, whose masterpiece *On the Origin of Species* still has a fundamental influence on biological thinking [2]. Darwin specified three foundational problems on which every evolutionary theory must be based: Variation, selection, and inheritance. The process of descent with modification needs some source able to produce variation in organisms. This variation must be reliably passed on to subsequent generations. Finally, variation will be maintained within a lineage if it confers an adaptive advantage.

However, as observed by Darwin himself, it is not natural selection that induces variation, but its role is limited to maintaining favorable variations and eliminating harmful ones [2] (p. 63). "Laws of variations" and "laws of heredity" are the most notable absentees within Darwin's long argument. "The laws governing inheritance are quite unknown; no one can say why a peculiarity in different individuals of the same species, or in individuals of different species, is sometimes inherited and sometimes not so; why the child often reverts in certain characters to its grandfather or grandmother or other more remote ancestor; [...]." [2] (p. 13).

"I have hitherto sometimes spoken as if the variations [...] had been due to chance. This, of course, is a wholly incorrect expression, but it serves to acknowledge plainly our ignorance of the cause of each particular variation." [2] (p. 102).

Anyway, Evolution (Natural Selection), Variation, and Heredity represent the three foundational issues from which are derived the genealogy of problems that have long affected evolutionary biological thought. These three issues represent the conceptual triad to be taken into account in any evolutionary theory [3,4].

Nevertheless, biological thinking did not address these foundational problems in the same way, and the phenomena and roles attributed to each element of the conceptual triad have always been different.

Within the Modern Synthesis (MS), particular emphasis was placed on the role of natural selection and genes, the latter being conceived both as unit of variation and unit of inheritance. In this framework, evolution was defined as a change in the frequency of an allele within a gene pool [5] or, in a simpler way, "biological evolution consists of changes in the genetic constitution of populations" [6] (p. 21). However, as said by Scott Gilbert, reported in [7], "The modern synthesis is remarkably good at modeling the survival of the fittest, but not good at modeling the arrival of the fittest" [7] (p. 281).

This is one of the reasons for dissatisfaction with the Modern Synthesis. Another reason is that MS has black-boxed all the processes and mechanisms that are between genotype and phenotype or between genetic mutation and natural selection [8]. As stated by one of the greatest critics of MS, Conrad Hal Waddington: "It has been clear at least since von Baer's day that a theory of evolution requires, as a fundamental part of it, some theory of development. Evolution is concerned with changes in animals, and it is impossible profitably to discuss changes in a system unless one has some picture of what the system is like. Since every aspect of an animal is a product of development, or rather is a temporary phase of a continuous process of development, a model of the nature of animal organization can only be given in developmental terms" [9] (p. 11).

Evolutionary Developmental Biology (EvoDevo) is the theoretical proposal able to unlock the black box of development, and also able to investigate the influence of development on evolution and the influence of evolution on development [10,11]. So, one of the aims of EvoDevo is to investigate the causal-mechanistic interactions which exist between the processes of individual development and the processes of evolutionary change [12], and to identify the generative mechanisms responsible for the variations presented by organisms (the arrival of the fittest). However, as we will see in Section 3, the conceptual framework of EvoDevo is anything but homogeneous. This conceptual framework is characterized by a set of core concepts, such as: Novelty [13,14]; homology [15,16]; genetic regulatory networks [17,18]; evolvability [19,20]; developmental bias [21]; modularity [22]; and others. Evodevo makes use of a combination of "[...] tools, techniques, and findings of molecular biology, anatomy, physiology, functional morphology, cell biology, embryology, developmental genetics, paleontology, comparative genomics, and population genetics, [...]" [23] (p. 1). It is composed of a variety of research programs, such as comparative embryology and morphology programs, evolutionary developmental genetics programs, experimental epigenetic programs, and theoretical and computational programs [12,24]. Finally, it is guided by several questions concerning the relationships between evolution, development, and environment [10,12]. For these reasons, this pluralistic framework has been rightly conceived as a "[...] loose conglomeration of research programs [...]" [25] (p. 265).

Anyway, EvoDevo represents a hard opponent for the MS. In fact, it was presented as: A paradigm [24,26]; a research program [27]; a revolution or revolutionary science [17,28]; or a new synthesis or a revolutionary synthesis [29,30]. These terms, far from being linguistic labels, represent theoretical and conceptual tools useful for a theoretical comparison between different perspectives; such as MS and EvoDevo. Nevertheless, the relationship between these two perspectives is far from clear. Is EvoDevo an expansion, an extension, or a revolution compared to the Modern Synthesis? EvoDevo "[...] is a discipline still in search of its identity" [27] (p. 213) whose conceptual and theoretical foundation has not yet been properly founded [27,31].

Section 2 describes the pluralistic framework of EvoDevo, with particular attention paid to how the various accounts conceive the processes of evolution and development.

In Section 3, we highlight the relationship between EvoDevo with the Extended Evolutionary Synthesis (EES) and Standard Evolutionary Theory (SET).

In the last section, we present some theoretical proposals to make EvoDevo take a few steps forwards.

2. One, No One, and Hundred Thousand EvoDevo

Vitangelo Moscarda is the protagonist of the novel 'One, no one, and hundred thousand', written by the Sicilian Nobel laureate Luigi Pirandello. In this novel, the protagonist realizes how his identity, contrary to what he believed, is not something certain, stable, and consistent for everyone, but each person has their own view of the personality and characteristics of Moscarda. The result is that his identity becomes questionable: From the certainty of a single initial identity to an identity distributed over a multiplicity of individuals, and finally to the impossibility of any form of identity. Although not afflicted by a comparable existential drama, we could say that the situation of the EvoDevo is similar, in some respects, to that of Vitangelo Moscarda.

2.1. One EvoDevo

Frequently, EvoDevo is defined as the study of how Evolution and Development are interrelated or, to put it another way, as the "evolution's influence on development and development's influence on evolution" [32] (p. 9). Moreover, how to conceive EvoDevo, that is, what kind of problems, themes, or aspects must be considered fundamental and worthy of investigation, "depends on whether the scientist is a developmental biologist, a paleontologist or an evolutionary biologist", or something else [30] (p. 75). Although it is a discipline that is more than 30 years old, EvoDevo, from a theoretical point of view, is (still) a discipline in search of its identity [27,32].

EvoDevo would like to be "[...] a synthesis of those processes operating during ontogeny with those operating between generations (during phylogeny)" [33] (p. xiii) (Hall&Olson 2003, p. xiii), and also an integration of several research areas such as genetics, ecology, paleontology, behavior, cognition, physiology, and so on [34] (p. 499). This framework is characterized by several core concepts (from life cycles evolution, modularity, plasticity, to evolvability, environmental induction, canalization, and more) [34] (p. 505), and by a set of problems agenda, whose purpose is to investigate the questions born at the interface between evolution, development (EvoDevo questions; DevoEvo questions), and also environment (EcoEvoDevo questions) [12].

So, we can say that the general aim of EvoDevo is "[...] to provide a mechanistic explanation of how developmental mechanisms have changed during evolution, and how these modifications are reflected in changes of organismal form", in such a way as to make possible "[...] to determine the mechanisms behind the 'arrival of the fittest'" [35] (p. 2). The problems begin when we enter in the "[...] incredibly intricate multi-level domains, mechanisms, processes, structures [...]" [36] (p. 259) that characterize development and start looking for the mechanisms responsible for the generation and variation of forms.

However, "the way we perceive the field today is often reflected in the way we reconstruct its history and, similarly, the way we reconstruct the history can reveal a lot about current assumptions" [37] (p. 2).

2.2. No One EvoDevo

A revolutionary event in the history of developmental genetics, so prominent that for some authors it represents the birth of the EvoDevo [17], has been the discovery of the homeobox [38,39] and of the role that this plays in the evolution and development of animal forms. The discovery was made in the 1980s. In the following years emerged what has been called the logic that controls development, based on a regulatory system in which a set of genes and transcription factors acts on other genes, activating or inhibiting them, in this way determining "[...] variations in the level, pattern, or timing of gene expression [...]" [40] (p. 579) which result in the construction of forms during development. This set of genes is highly conserved in the course of evolution, and it is held responsible for the construction of body plans and different body structures [17].

The discovery of the homeobox has been described as the Rosetta Stone of developmental genetics; on this foundation, it would be possible to achieve an understanding of the "[...] deep unity about pattern formation […] which underlies the apparent superficial complexity and diversity of animal development" [41] (p. 365). Both homology and evolutionary novelty are thus explained in terms of modifications based on the combinatory logic of activation/inhibition of the cis-regulatory systems; what has been called by Gould "Hoxology" [42].

The central role played by this gene regulation system, both in evolution and development, is expressed by what we may call Davidson's Syllogism: "Since the morphological features of an animal are the product of its developmental process, and since the developmental process in each animal is encoded in its species-specific regulatory genome, then change in animal form during evolution is the consequence of change in genomic regulatory programs for development" [43] (p. 27–28).

What characterizes the Hoxological approach is an explicit gene-centrism, according to which the great variety of animal forms is encoded in the combinatory logic of the cis-regulatory system, conceived as "[…] the essence of animal development" [44] (p. 25).

What is more interesting is that, according to some exponents of this approach [17,45], the EvoDevo/Hoxology represents a revolutionary expansion of Modern Synthesis; a third act (the first one was the Darwinian theorization, the second one was the MS) that complemented the MS with a hitherto missing piece: A mechanistic explanation for the evolution and development of phenotypes. As stated in [45], both phyletic and morphological evolution are characterized by the same genetic and developmental mechanisms, even suggesting an equation between Macroevolution = Microevolution. If so, EvoDevo would not represent "[…] any fundamental conceptual problem for evolutionary biology" [46] (p. 386); rather, it would be the completion of the conceptual framework of MS through the addition of causal mechanisms able to explain the origins of variation. However, this perspective was accepted with both cautious optimism [47] and skepticism [48] by exponents of the Standard Evolutionary Theory.

But as pointed out by [37], "[…] evolutionary developmental biology is not as uniform as the image of a new scientific discipline and the powerful icons of "the genetic toolkit for development" and the almost magical qualities of "Hox genes" seems to suggest." (p. 1). The study of the evolution and development of the phenotype "[…] requires thinking both within and outside the paradigm of transcription encoding factors" [49] (p. 129). So, EvoDevo must be something more than a "mere" Hoxology.

2.3. One Hundred Thousand EvoDevo

Marta Linde Medina introduces a distinction between "two versions of EvoDevo" [50], which reflects internalist and externalist traditions—as presented by Ronald Amundson [51], and which represents two ways of conceptualizing evolution and development. One version, EvoDevo1, is linked to the discovery of Hox genes in the 1980s [52,53], and the consequent progress of developmental genetics [17,43]. This EvoDevo perspective is closely based on a gene-centric conception of both evolution and development, in which are often used metaphors such as those of the genetic program and of genes as depositories of ontogenetic and phylogenetic information. The development of organismal form is based on switches, regulatory sequences, transcription factors that make up a complex genetic regulatory network on which depends the construction, as well as the evolution, of body plans. The Natural Selection will act on variations in the wiring of the networks able to produce both microevolutionary and macroevolutionary phenomena. Thus, we are faced with a "perfect" synthesis of mechanisms acting during ontogeny and the mechanism of natural selection [17,43], in harmony with the externalist tradition. The other version, EvoDevo2, is linked to an internalist perspective, in which a fundamental role is recognized to the inherent properties of living matter [54], the self-organizing phenomena and events [55], and the overall set of interactions that occur during ontogeny and across several levels of organization [56]. In this perspective, the genetic circuitry, even if it plays an important role, is conceived as a post-hoc effect, i.e., a stabilizing factor of forms, rather than an exclusive generating factor. This last perspective is seen as "[…] a subdiscipline of an extended evolutionary synthesis (EES) […]" [50] (p. 9). On the other hand, the version presented by [17], for example, is more a genetic theory of morphological evolution [46], and therefore it is only one component, one aspect, of a more general theory of evolution where development is finally and adequately taken into consideration.

This first distinction is useful because it allows making a coarse theoretical distinction between two alternative conceptions of EvoDevo [25]: EvoDevo in a narrow sense, including mainly Hoxology; and EvoDevo in a broad sense, including a wider set of perspectives. This distinction can be further developed into a typological classification of positions, based on [57].

Callebaut, Müller, and Newman [57] identify various accounts of Evo-Devo or, as they call them, twelve evo-devo packages, "[…]which in part reinforce one another, but in part also compete" (p. 34). These accounts are the result of the combination of different theoretical, methodological, and epistemological perspectives. We consider it useful to present an overview of these accounts to highlight the plurality of perspectives that characterizes EvoDevo. In the end, we will make some general remarks.

Core EvoDevo (1) is described as a perspective "[…] encompassing both the program to explain evolution through changes in development and the program to reframe evolutionary biology more drastically along developmental lines" [57] (p. 35).

Gene Regulatory Evolution (2) basically corresponds to the Hoxology presented in the previous section.

Epigenetic EvoDevo (3) has as its main target morphological evolution; notably the processes and mechanisms that characterize the generation, fixation, and variation of structural building elements of organisms [58]. Within this perspective, a further problem is represented by the genotype–phenotype relation, that is, how genes and morphologies are related. On the one hand, genetic and developmental pathways may change during phylogeny while morphologies remain relatively constant; on the other hand, similar gene expression patterns may correspond to different morphologies [15]. In order to describe the processes of form generation, it is not enough to give a description of the genetic control systems: Our account must include "[…] the generic material properties of, and interaction dynamic among, cells, tissues and their environments"[59] (p. 66). The latter factors constitute the epigenetic properties of development, such as: "(i) interactions of cell metabolism with physicochemical environment within and external to the organism; (ii) interactions of tissues masses with the physical environment on the basis of physical laws inherent to condensed materials, and (iii) interaction among tissue themselves, according to an evolving set of rules" [56] (pp. 305–306). The relationship between genes and forms is indirect, and epigenetic mechanisms represent the principal factor directing morphological evolution; meanwhile genetic networks are co-opted subsequently to play a stabilizing role rather than a generative one. In the Epigenetic account, homology has an important role, insofar as the main clue of a common descent and of organismic evolution. Indeed, homologues represent attractors of morphological design, i.e., "autonomous organizer of the phenotype in an evolutionary lineage" [59] (p. 70–71), and that during the phylogeny provides the basis for further modifications. Therefore, there is an attempt to recognize a relatively autonomous role of the morphological level with respect to the underlying mechanisms, especially the genetic ones.

Process Structuralism (4) "[…] assumes that there is a logical order to the biological realm and that organisms are generated according to rational dynamic principles" [60] (p. 91). So, a theory of evolution must include a theory of biological form, which explains how organisms are structured and by which transformations are characterized. This account puts center stage organisms conceived as "dynamically transforming systems" described as fields, i.e., "[…] domains of spatial order, defined by internal relationships, that change in time according to well-defined principles or rules" [61] (p. 129), and moves against the Neo-Darwinian assumptions according to which, to explain evolutionary phenomena, it is enough the action of natural selection on random genetic mutations [62]. So, what is proposed is a theory of morphogenetic fields defined as complex dynamic systems in which "[…] genetic and environmental factors determine parametric values in the equations describing the field and therefore act to select or stabilize one manifest form from the set of forms which are possible for that type of field" [61] (p. 121). In this perspective, the patterns of gene expression alone cannot provide neither a causal explanation of the developmental processes nor an understanding of the generative principles underlying the development and evolution of forms. On the contrary, development is conceived as a hierarchical process along which various factors can alter the dynamics of the system. In this way, it is possible to define the space of possible biological forms (generic forms), i.e., "the natural kinds that are revealed in evolution, the basic biological forms that are all transformations of one another under changes in the detailed dynamics of morphogenetic fields" [61] (p. 130).

The morphogenetic field represents the generative unit both in development and evolution. Evolution is conceived as a "time-dependent exploration of a set of possibilities under internal (genetic) and external (environmental) parametric variation" [60] (p. 96); these possibilities are given by "generic states of forms [. . .] whose distribution in the space of developmental trajectories defines the set of possible forms" (p. 96). What emerges is a perspective on evolution and development in which: "The hierarchical nature of this generative process leads naturally to a hierarchical taxonomy of biological forms: ontogenesis provides the logical foundation for understanding phylogenesis" [60] (p. 130).

Self-Organization of Biological Complexity (5) explains the complexity of the organization and forms of biological systems as due to intrinsic dynamics and patterns, rather than to the external action of natural selection. This account, as was Process Structuralism, was inspired by the tradition of Rational Morphology (Goethe, Cuvier, Geoffroy Saint-Hilaire, Owen, and more recently, D'Arcy Thompson), according to which "[. . .] organisms were built up by combinatorial variations of a small number of principles"[55] (p. 4). The aim of Rational Morphology, as well as of the Self-Organization account, is to discover the laws of form underlying biological organization. We can define Self-Organization as "[. . .] a process in which pattern emerges at the global (collective) level through interactions among the components of the system at the individual level, without these interactions explicitly specifying the global pattern [. . .]" [63] (p. 79).

This means that the order performed by the organisms is not due to the action of natural selection, which does not have the ability to generate order, but to the inherent properties of complex systems which constitute the organisms themselves [55]. The main processes occurring during development, endowed with self-organizing properties, are cellular differentiation and morphogenesis. Cellular differentiation is the result of the dynamical behaviors of genetic regulatory networks, in which cell types represent "[. . .] a recurrent pattern of gene activity" [55] (p. 442). By contrast, morphogenesis is defined as "[. . .] the consequence in time and space of the structural and catalytic proprieties of proteins encoded in time and space by the genome, acting in concert with nonprotein materials and with physical and chemical forces to yield reliable forms" [55] (p. 410). Within this account, a significant problem is represented by the relationship between natural selection and self-organization. There are different possible hypotheses: A primacy of natural selection over self-organization; self-organization as a constraint or an auxiliary of natural selection; natural selection and self-organization as two sides of the same coin [64]. The last hypothesis is suggested by Kauffman: "Selection achieves and maintains complex systems poised on the boundary, or edge, between order and chaos. This systems are best able to coordinate complex tasks and evolve in a complex environment. The typical, or generic, properties of such poised systems emerge as potential ahistorical universals in biology" [55] (p. xv).

Structural Modeling (6) is a computational and theoretical approach to the study of RNA folding, that can be presented as "minimal model of a genotype-phenotype relation" [65] (p. 1164), but it cannot represent the entire process of organismal development. Through this model, it is possible to explore what kind of molecular phenotypes (RNA secondary structure, or RNA shape) can be reached from gradual modifications of the genotype. Moreover, it is possible to understand how genotype and phenotype influence each other, that is, how particular phenotypes are produced from particular genotypes, how these phenotypes are targets of selection, and, finally, how this selection feeds back on the genotypes and developmental mechanisms producing the initial phenotypes. The evolution of the phenotype depends on the topological structure of the phenotypic space as result of the genotype–phenotype map. What emerges from the simulations is that the possible phenotypes accessible from various genotype sequences are limited; there is a redundant mapping that maps many genotypes onto a single phenotype [66]. By gradually varying the sequences of the genotype, it is possible to observe the result on the phenotypic space. Many variations will be neutral, that is, will not lead to a significant change in the process of folding of RNA and, consequently, in the final shape of RNA. So, a Neutral Network can be defined as "[…] a mutationally connected set of genotypes that map to the same phenotype" [66] (pp. 79–80). Therefore, it is possible to distinguish between continuous and discontinuous phenotypic changes: The former results from the exploration within the boundary of a neutral network, while the latter results from crossing the boundary between two adjacent neutral networks, mapping onto two different phenotypes. Although this perspective does not provide a general representation of the development of organisms, it allows us to study, in a limited and controlled manner, phenomena such as epistasis, phenotypic plasticity, constraints on variation, canalization, modularity, phenotypic robustness, and evolvability, whose large-scale study would require considering a large and complex set of factors and variables such as to make investigation intractable.

Dialectical Account (7) rejects the ideology of biological/genetic determinism, which reduces properties and characteristics of complex biological systems to the organization and processes of their simplest constituent components (genes), to which causal and ontological priority is also attributed [67,68]. Conversely: "Dialectical explanations, on the contrary, do not abstract properties of parts in isolation from their associations in wholes but see the properties of parts as arising out of their associations. […] according to the dialectical view, the properties of parts and wholes codetermine each other" [68] (p. 11).

This means that in the explanation of development, there is no place for genetic determinism or program metaphors; neither for a naive dualistic conception according to which some traits are caused by genes while the environment plays the role of filling the remaining causal gaps. On the contrary, development is a process in which there is constant interaction between genes and environment where "[…] random variation in growth and division of cell during development […]" represents an important source of variation (developmental noise) during ontogeny [67,69]. Moreover, the environment is not conceived as something external and independent from organisms and to which organisms must adapt, but environment and organism are engaged in a process of co-determination [69] in which: "[…] organisms determine which elements of external world are put together to make their environments and what are the relations among the elements that are relevant to them" [69] (p. 51); moreover, "[…] organisms not only determine what aspects of the outside world are relevant to them by peculiarities of their shape and metabolism, but they actively construct […] a world around themselves" (p. 54); and finally, "[…] organisms […] are in a constant process of altering their environment" (p. 55). In this account, both evolution and development must be conceived in a more complex causal perspective, according to which "[…] the genome, the proteome, the traitome, the behaviorome, and the societome […]" [70] (p. 334), and the environment in general, are characterized by complex nonlinear and bidirectional dynamics.

Systems Biology (8) is an emerging approach of which it is difficult to provide a comprehensive definition; within this field are present various traditions, such as a physiological [71] and a computational one [72]. Anyway, to provide a general definition, systems biology "is the study of the behavior of complex biological organization and processes in terms of the molecular constituents" [73] (p. 504). The aim is to obtain a "system-level understanding of biological systems" [72], and to achieve this goal, it is necessary to acquire and integrate many kinds of data; especially from the '-omics' disciplines [74]. From the analysis of these complex biological networks, it may be possible to understand the interaction between DNA, phenotype, and environment [71]. The principal aim of this approach is a multi-level integration [75]: "Identifying all genes and proteins in an organism is like listing all the parts of an airplane. While such a list provides a catalog of the individual components, by itself it is not sufficient to understand the complexity underlying the engineered object" [76] (p. 1622).

This means that systems biology moves beyond the single level of analysis method to adopt an integrative perspective, including both top-down and bottom-up approaches. Likewise, exclusive knowledge of genetic regulatory networks and biochemical interaction is not enough to explain the organization of organisms. What must be pursued is an understanding of the organizing principles and the behaviors of the various levels of hierarchical organization. In this sense, systems biology looks for "[…] multiscale, multilevel explanations of organismal properties" [77] (p. 8), in which there is no privileged level of explanation to which to reduce biological phenomena. The complex framework presented by systems biology also forces us to reconsider the causal relationships among levels of organization. Namely, there is not only upward causation as in reductionist models, from genes to organismal behavior, but between levels of organization there are feedback circuits through which higher levels constrain the behavior of the lower levels [78]. This perspective requires a rethinking of development, which cannot be conceived of as a simple, linear, and unidirectional process directed by a program or a single level of organization, but as a complex system of interactions; and afterward a rethinking of the factors and aspects that are relevant in heredity and evolution [75,79].

Cybernetic Synthesis (9) is based on Jean Piaget's conception of evolution, closely based on the role of phenotypes and behaviors as main actors of evolutionary phenomena [80]. In his conception, behaviors are viewed not only as a result of evolution, but also as one of its determinants [80] (p. xi). Piaget's theory of organismic development and evolution is influenced by Paul Weiss and Conrad Waddington. If we want to understand the role of behavior in evolution, we must consider the hierarchical system in which the organisms are located. If we conceive the organism as a system of concentric shells, in which the innermost corresponds to genes while the outermost corresponds to the environment. What characterizes organism's dynamics is a complex network of bidirectional relations [81]. Thus, organisms are characterized by a system of cybernetic interactions, in which a single level (e.g., genetic) cannot cause or control, unilaterally and simply, the highest levels (ex. morphology and behavior). Rather, the highest levels can feed back, altering the dynamics of the lower levels.

Therefore, behavior is conceived "[…] as the expression of the overall dynamics of organization in its interaction with environment and as a source of supersessions and innovations for as long as the environment or environments continue to contain any elements creating obstacles for the organism" [80] (pp. 140–141). In this way, the behavior and other environmental factors can stress a response (in the hierarchical system) of the organism to produce a modified phenotype (phenocopy), which will be stabilized by the genetic level in successive generations [80] (p. 73–83). "The Organism is an open system, a necessary precondition of whose functioning is behavior; and […] it is of the essence of behavior that it is forever attempting to transcend itself and that it thus supplies evolution with its principal motor" [80] (p. 139).

So, according to Piaget, behavior can be considered responsible "[…] for the far-reaching morphogenetic changes of macroevolution" [80] (p. 140).

Reproducer Perspective (10) represents a radical way to think about a unifying theory of heredity, development, and evolution "[…] that accounts for the way heredity and development are entwined in reproduction processes" [82] (p. 253). Units of heredity and units of development are intertwined in units of reproduction [83], and the latter, the reproducers, are the units of evolutionary transition. Reproducers are defined as entities that have the capacity to make more reproducers, i.e., [84] development is presented as "[…] the recursive acquisition (over a compositional hierarchy of parts and wholes) of a capacity of reproduce" [85] (p. 187). In other words, "[…] reproduction is the recursive propagation of the organized capacity to develop and development is the ordered realization of the capacity to reproduce" [86] (p. 267).

In this processual perspective, development is a complex process that does not concern a single level of organization, but involves an organized mechanism of reproduction, spanning over several levels. Meanwhile, evolution is conceived as a "[…] descent with modification of a population of reproducers" [85] (p. 187).

Developmental Systems Perspective (11) is defined as "[…] a general theoretical perspective on development, heredity and evolution" [87] (p. 1), representing an attempt to go beyond classical dichotomies such as nature–nurture, genes–environment, and gene-centric approaches. In this perspective, the fundamental unit is the developmental system conceived "[…] as a complex of interacting influences, some inside the organism's skin, some external to it, and including its ecological niche in all its spatial and temporal aspects, many of which are typically passed on in reproduction […]. It is in this ontogenetic crucible that form appears and is transformed […]" [88] (p. 39). The stability and variation of forms are not explained by the presence of a genetic program, however complex it may be, but by the contingent process of construction that involves several factors during ontogeny.

Developmental Systems Theory has been regarded as a key partner of EvoDevo, although problematic [3,89]. The major tenets of DST present a serious challenge to traditional evolutionary concepts. This includes [88] (p. 2): (i) The presence of multiple causes involved in developmental processes; (ii) the contingency of development processes towards the context and the other states of the system; (iii) an extended conception of heredity; (iv) a conception of development as a construction (there is not a preconceived plan); (v) there is not a privileged level of control, but this is distributed over multiple factors; (vi) evolution is conceived as changes in the system of interactions between organism and environment.

So, Development Systems Theory conceives "[…] both development and evolution as processes of construction and reconstruction in which heterogeneous resources are contingently but more or less reliably reassembled for each life cycle" [88] (p. 1).

Core Configuration Model (12) represents the attempt to sketch an evo-devo perspective to human cognition, able to explain, in particular, the behaviors and organization of social cognition. Here too, the Neo-Darwinian dichotomy between genes and environment and the gene-centric perspective are refuted and replaced by the notion of recurrence or repeated assembly, conceived as "recurrent entity-environment relations composed of hierarchically organized, heterogeneous components (which may themselves be repeated assemblies) having differing temporal scales and cycles of replication" [90] (p. 59). Biological entities present a hierarchical organization, spanning from DNA, cells, and tissues to organisms, groups, and populations. This hierarchical organization implies that organisms are "[…] the contingently developmental result of various genetic and epigenetic resources […]" [90] (p. 59), and their traits are "the result of a relational linkage, contingently and concretely situated in a specific context of interacting elements spanning between the various level of hierarchical organization characterizing biological entities" [91] (pp. 278–279). In this way, the elements found on a level of organization represent the context for elements of another level. All the elements which constitute that hierarchical system are characterized by both upward and downward causation [91] (p. 278).

Thus, we can define repeated assembly as recurrent constructive relations composed of heterogeneous and hierarchical components. Therefore, development is a non-programmed process, which emerges from the interactions among various resources that must be repeatedly assembled over generations and which must be inherited in a stable manner [91] (p. 279).

2.4. The Search for Unity

This overview of EvoDevo approaches, based on the typological classification presented in [57], is not necessarily complete. In fact, it is possible to identify further theoretical perspectives, which we present here in a concise manner, but whose complexity would require a broader discussion. The **Developmental Plasticity perspective (13)** is an account that considers the role of developmental plasticity as a leader, not necessarily follower, in phenotypic evolution. The latter is conceived more as a complex reorganizational phenomenon rather than due to a genetic mutation [92,93]. The **Ontophylogenetic perspective (14)** represents a radical holistic account in which evolution and development "[…] are the two inseparable sides of a single reality produced by a unique process" [94] (p. 5). These two mechanisms are joined in a genealogical line (lignée généalogique) that is the result of a continuous production process of organisms. The unit of analysis of phylogeny, i.e., the species, and the unit of analysis of ontogeny, i.e., the individual, are two aspects of a continuous phenomenon that encompasses both: Ontophylogenesis [95]. The **Probabilistic Epigenesis perspective (15)** represents a meta-theoretical model of development which emphasizes the role of developmental manifold, conceived as a set of bidirectional influences within and between levels of organization/analysis (including genetic activity, neural activity, behavior, and physical, social and cultural environment) [96,97]. The **Eco-Evo-Devo perspective (16)** is a complex approach in which "[…] the environment is a source and inducer of genotypic and phenotypic variation at multiple levels of biological organization, while development act as regulator that can mask, release, or create new combination of variation" [98] (p. 107). This perspective aims to discover "[…] the rules by which an organism's genes, environment, and development interact to create the variation and selective pressures needed for evolution" [99] (p. XIV). The Behavioral/Cognitive perspective (17) investigates the development of behavior and cognition and how these two phenomena can contribute to modify and determine the dynamics of development and evolution [100,101]. The **Character Identity Networks perspective (17)** is a genetic approach to morphological evolution, which considers the role of genes and genetic networks in determining the identity and variational modalities of a morphological character, also taking into account the developmental context (species-specific mode of development) [18].

However, this does not mean that it is possible to indefinitely extend the set of theoretical proposals related to EvoDevo. The risk is to take an inflationary attitude that identifies more differences in theoretical proposals than necessary. For this reason, it is appropriate to associate a deflationary attitude, which, instead of multiplying the possible perspectives, brings them together, smoothing out the differences. For example, it is possible to combine in a single perspective those proposals **(9, 12, 17)** that focus on the role of behavior and cognition in the light of ethology (Bateson [100]) and current cognitive theories (Stotz [101]). Moreover, it is possible to amend those Hoxological perspectives by taking into consideration the fundamental role of genetic networks within a broader and more complex developmental context, as in [18,102]. The various accounts presented previously represent theoretical perspectives, that is, particular points of view from which it is possible to investigate evolution and development [103,104]. Each perspective focuses on a particular aspect of the phenomena investigated, highlighting certain characteristics, and omitting others. So, the creation of a typology is a necessary step that must be followed by further elaboration and integration of the various perspectives to develop a general conceptual framework of EvoDevo. This is a complex task of theoretical synthesis, which cannot be done here.

Moreover, it is possible to identify another issue in this typology, namely a theoretical categorization problem, according to which (as pointed out opportunely by one reviewer) different perspectives (specifically from **4 to 10**) have been formulated outside the theoretical framework of EvoDevo and sometimes well before it emerged. If by EvoDevo perspectives we have to mean only those that explicitly refer to the same conceptual apparatus, then the solution to this problem would be trivial: It would be enough to identify all the theoretical proposals that use concepts like homology, novelty, genetic regulatory systems, modularity, and so on, to identify which can be labeled as EvoDevo and which cannot. But if the outsider perspectives contribute to a greater understanding of the processes, phenomena, and factors that characterize evolution and development through introducing new theoretical tools, then why not integrate these new tools within EvoDevo's theoretical apparatus? For example, Self-organization is not a phenomenon that can be found only in the biological field, but also in the physical and chemical fields. Therefore, even if it is a concept that has not been formulated within EvoDevo, it can be a useful tool to understand and describe the development of organisms as a phenomenon that emerges from the organizational complexity of the organisms themselves.

In summary, we consider this typology useful because: (i) It makes explicit the pluralistic framework, or the conglomerate structure, [25] that characterizes EvoDevo; (ii) it allows us to isolate various proposals and positions concerning the three fundamental processes (evolution, development, and heredity); (iii) it provides a conceptual basis for comparing different theoretical proposals with each other; and (iv) it sets a task for the future, i.e., "[…] to probe how deeply entrenched the metacommitments of the advocates of the various perspectives on evo-devo really are" [57] (p. 40).

Inspired by Robert [105] (p. 596), we can graphically represent the various perspectives as points distributed on a continuum (Figure 1). We can define a criterion according to which the various positions can be arranged along this continuum according to their degree of inclusiveness. A criterion that takes into account which factors, mechanisms, and processes are involved in explaining and describing evolution and development. Simplifying, we could say that in this theoretical continuum, we can see reductionist to the right and holist to the *left* [106] (p. 41). On the right side, there are those approaches that reduce development processes to a single level and are based on notions of information, regulation, and control. On the left side, there are those perspectives that have a more global, dynamic, and inclusive approach.

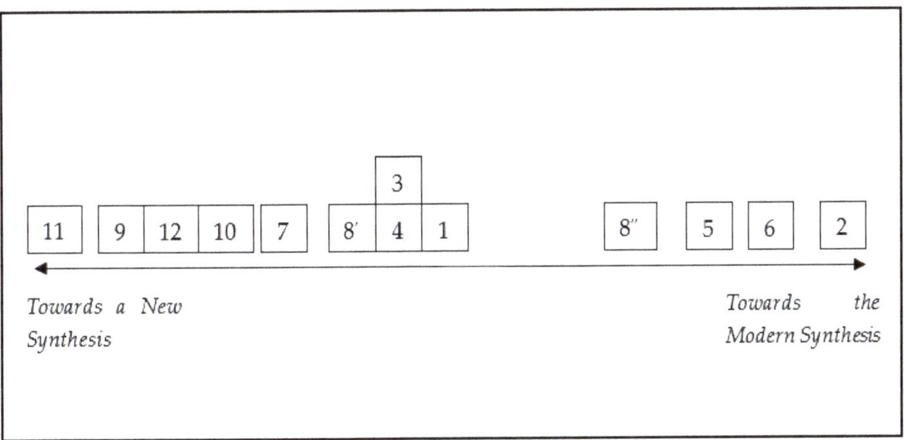

Figure 1. Theoretical Continuum of the various EvoDevo perspectives.

At the right end of the continuum, there is Gene Regulatory Evolution/Hoxology, with its gene-centric and reductionistic approach that presents the DNA/genetic regulatory system as depositary of the instruction to build the organisms, the only hereditary factor is represented by genes, and morphological evolution is due to modification in the regulatory system. This EvoDevo account is closer than the others to the theoretical framework of the MS. At the left end, there is Developmental Systems Theory, conceived not only as a synthesis of evolution, development, and heredity, but also as a radical rethinking of evolutionary theory. In the middle we find the core EvoDevo, guiding the theoretical enterprise and setting the concepts and the problems that have to be addressed. All other accounts are distributed throughout the continuum, according to the criterion of inclusiveness.

This theoretical continuum is an artificial, simplistic, and limited way to represent the relationships between the various perspectives. It represents a theoretical exercise and with obvious limitations; for example, it is not possible to define a metric to quantify the distance between theoretical perspectives. However, we believe that it can be a useful visual tool to understand how some perspectives are more conceptually closer than others, and an initial basis for further and more complex processing.

So, because of this pluralistic framework, it is difficult to conceive EvoDevo as a unitary theory; rather, it seems to be "[…] the pursuit of specific epistemic goals, such as the explanation of evolvability, evolutionary novelty, or homology" [107] (p. 396), or a Trading Zone, i.e., "[…] a variety of disciplines, styles, and paradigms negotiating heavily with one another" [108] (p. 459). It is thanks to this pluralistic framework that EvoDevo plays an ambiguous role within evolutionary thinking.

At times, it seems to be in continuity with Modern Synthesis, while at other times a challenge. In the next section, we will investigate the role that EvoDevo plays within evolutionary theory.

3. A Synthesis within a Synthesis

In the last few years, there has been a growing interest in a new evolutionary synthesis, not conceived however as a paradigm shift, or revolution, but as a "[...] series of complex developments that build on its three major predecessors: Darwinism, Neo-Darwinism, and the MS itself" [109] (p. 2748). This Extended Synthesis represents a multifaceted research program expanding the conceptual and theoretical boundaries of evolutionary theory [110].

If EvoDevo has been defined as a synthesis of evolutionary and developmental biology, that it is more than "[...] developmental biology grafted onto evolutionary biology [...]" [27] (p. 214) or a modern synthesis plus EvoDevo [111] (p. 190), so, what is the role or status of EvoDevo within the general framework of evolutionary theory?

In 2014, an article was published in which several of biologists and philosophers of biology called for a rethinking of evolutionary theory (the Reformists) [112] contrasted others who did not consider it necessary (the Conservatives) [113]. The first group of scientists advocates of an Evolutionary Extended Synthesis (EES), while the second group has a perspective in continuity with MS, a Standard Evolutionary Theory (SET). The role played by EvoDevo within each framework is different.

EES represents an "alternative vision of evolution" organized around a broader framework with its structure, assumptions and predictions, which "[...] will shed new light on how evolution works" [112] (pp. 161–162). The Reformists present four core processes neglected by SET (developmental bias, plasticity, niche construction, extra-genetic inheritance), that represent the main weapons in what is "[...] a struggle for the very soul of the discipline" (evolutionary biology). The fundamental message that emerges from this skirmish is a notion of development as "[...] a direct cause of why and how adaptation and speciation occurs, and of the rates and patterns of evolutionary change" (p. 164), and also the view "[...] that variation is not random, that there is more to inheritance than genes, and that there are multiple routes to the fit between organisms and environments" (p. 164). Based on these assumptions, the EES represents "[...] a different framework for understanding evolution [...]" [114] (p. 3) or "[...] a comprehensive new synthesis" [115] (p. 8).

Meanwhile, on the Conservatives side, all is well: Adaptation and speciation are "[...] two of the most fundamental evolutionary processes"; DNA is "[...] the material basis for heredity and trait variation"; the putatively neglected processes were known since Darwin, and "[...] are already well integrated into evolutionary biology" [113] (p. 163). Then, more than a rethinking of evolutionary theory, the conservatives are more willing to a modest expansion covering the supposed neglected phenomena. However, the central tenets of evolutionary biology remain valid [113].

Futuyma [47] further discusses these aspects. The SET has already expanded its contents to include "[...] transposable elements, exon shuffling and chimeric genes, gene duplication and gene families, whole-genome duplication, de novo genes, gene regulatory networks, intragenomic conflict, kin selection, multilevel selection, phenotypic plasticity, maternal effects, morphological integration, evolvability, coevolution and more [...]" [47] (p. 2). This expresses the inclusive character of the ES, which adopted its heuristics to include and expand the range of phenomena it was able to explain. More interesting, according to Futuyma, EvoDevo "[...] described by some adherents to an EES is oddly different from the research literature that has made the most substantial progress in evolutionary developmental biology" [47] (p. 8). So, Futuyma endorses the hoxological approaches [Carroll, Davidson], stating that "mechanistic understanding of gene action, of regulatory circuits, of the conservation of elements in the "genetic toolkit," and their association with different downstream genes are rapidly deepening our understanding of evolutionary changes in form"[116] (p. 54). According to this, EvoDevo/Hoxology implement and complete the ST with mechanical causes able to link macroevolution to microevolution.

On the Reformist side, the theoretical tenets that characterize Extended Evolutionary Synthesis [114] are organized around a structure and a series of assumptions and predictions. In this perspective, EvoDevo plays a key role in shaping or extending the Synthesis [109]. EvoDevo not only "[...] provides a causal-mechanistic understanding of evolution [...]" (p. 3) involving concepts such as developmental bias, evolvability, phenotypic plasticity, or facilitated variation; but also affects the overall structure of the synthesis, with a hierarchical, constructive, decentered (from the genetic level), and causally complex perspective on development and evolution [115]. In the clash between these two paradigms, Evodevo assumes an ambiguous position: One side refers to EvoDevo in a narrow sense, as Hoxology; the other one takes a broader perspective.

On the Synthesis and Expansion of Theories

We can say that EvoDevo, depending on the framework we adopt, is (i) a synthesis (desired, but not yet achieved) between developmental and evolutionary biology; (ii) a piece of a more complete synthesis, based on the MS; (iii) a leading (but ancillary, in respect to the overall theory) component in a new Extended Evolutionary Synthesis. We believe there is the risk that "[...] the boundaries of evolutionary developmental biology are becoming blurred and harder to demarcate, and so its goals and questions" [117] (p. 178).

About the term synthesis we can move from the following sketch [118]: "(i) a synthesis is a unification of originally disparate scientific structures (models, sets of models, theories, or even disciplines), and (ii) in the synthesized structure there is epistemic parity between the structure so unified" (p. 1217).

We can further distinguish between disciplinary integration, disciplinary synthesis, conceptual integration, and conceptual synthesis [119]. We have a disciplinary integration when two disciplines are compared, and "the result is one disciplinary structure that integrates two previously distinct disciplines" in which "[...] the individuality of the original parts is lost or effaced" [119] (pp. 311–312). A conceptual synthesis is when "[...] the significance of concepts in one discipline is evaluated for another (and vice versa) [...]" with the "[...] blending of one or more parts to produce a new entity where the individuality of the original parts is not dissolved, though potentially transformed" [119] (pp. 311–312). In the meantime, a disciplinary synthesis "[...] produces a new discipline without dissolving those from which it was synthesized[...]" [119] (p. 312). Finally, a conceptual integration "[...] refers to how more than one concept can be merged into a single new concept for various purposes" [119] (p. 312). EvoDevo has been conceived, perhaps too rigidly, as a disciplinary synthesis.

Another useful distinction is made by James Griesemer [120], who distinguishes between different modes in which we can conceive a synthesis (in particular, the EES). The first mode is Theory Extension, which is a conceptual theory-revision or theory-building that "[...] works by building out from (and maybe modifying) a core theory to incorporate new theoretical principles (perhaps from another theory)" [120] (p. 323). The second mode is Domain Expansion, which consists of adding new phenomena "[...] under the umbrella of a theory and guidance of its perspective" [120] (p. 324). Lastly, Practice Integration consists of incorporating "[...]a practice into a workflow from other lines of work, specialties, disciplines, possibly changing what work is produced or changing interpretations" [120] (p. 324). Obviously, these three types of synthesis can also be combined with each other, but what is important in a process of theory extension, such as EES, is that "[...] the mode (addition, subtraction, modification, recombination) and target (concept, principle, model, perspective) must be specified" [120] (p. 323).

Let's go back to EvoDevo and the two evolutionary frameworks. EvoDevo/Hoxology represents mainly an instance of Practice Integration, because it adds to MS/SET new (old) conceptual tools and empirical methodologies, i.e., the possibility to describe the evolution and development of morphologies through concepts like homology, novelty, body plans, etc., and explain experimentally the processes of construction of forms with the tools of developmental genetics. From a theoretical point of view, it is not required a substantial modification of Hoxology and SET, rather a gradual process of accommodation and conceptual synthesis, for example, with population thinking.

On the contrary, the EES represents at the same time a domain extension, since new phenomena and mechanisms are added to explain the processes affecting evolution and development, and a theory extension, because it requires a conceptual revision of how it conceives evolutionary theory. However, it does not represent a revolution of the old conceptual structure of MS, but an alternative research program [117] (p. 173).

In conclusion, EvoDevo conceived as a synthesis could represent a radical rethinking of the evolutionary theory [121], a kind of theory-revision/theory-building. But there are several conceptual difficulties. One of these concerns the pluralist framework that characterizes EvoDevo, and which represents an obstacle to conceptual and disciplinary synthesis. Another conceptual difficulty concerns the nature of development itself, and the possibility to establish a theory of development [122].

We agree with [24], that the main challenge of EvoDevo is the integration/synthesis between the various approaches, perspectives, or accounts presented above. The greatest contribution of EvoDevo will not consist in an expansion of the empirical basis, but in a new theoretical framework able to coherently hold together a variety of processes, mechanisms, phenomena, levels of description, phenomenology, and ontologies [123]. So "[…]a detailed epistemological analysis, […] of these programs and their underlying assumptions is therefore crucial for the future of evo-devo as a synthetic enterprise" [24] (p. 359).

What is needed is a conceptual synthesis able "[…] to provide a conceptual foundation for different research programs that will ultimately explore the fuzzy edges and areas beyond the core of the synthesis, then we advocate an open conception of synthesis rather than a closed view based on integration of existing paradigms" [24] (p. 359). We think that this synthesis must be based on a theory that takes development seriously [3]. The problem is that we don't "[…] have a sense of a general theory or principle of what development is beyond the slogans of decades or centuries past" [120] (p. 321).

4. Conclusions: Towards a Theory of What?

Although EvoDevo has been defined as "[…] evolution's influence on development and development's influence on evolution" [32] (p.9), little attention has been devoted to the formulation of a theory of development that could be integrated within the EvoDevo Synthesis. We agree with [124] "[…] that current skepticism partly results from a failure to articulate evo-devo's conceptual foundation properly" (p. 100), and in particular its developmental foundations. As stated by [31], "compared to evolutionary biology, developmental biology has much less elaborated theoretical foundations to the extent that its subject, development is rarely defined, and usually delimited in purely operational terms […]" (p. 3). From this follows that "[…] a revised (more flexible and more comprehensive) concept of development will require a strong revisitation of evo-devo's research agenda" [31] (p. 3).

The real innovative aspect of EvoDevo is precisely its focus on development. Without an adequate theory of development, at best, EvoDevo "shapes the extended synthesis" [11] as in EES, at worst, it can be reduced to a more articulated version of developmental genetics. In the latter case, the EvoDevo, going from the main antagonist of the MS, is likely to become its greatest ally. In the former case, without a theory of development, EvoDevo would not exploit its full potential.

What does it mean to develop a theory of development? Elaborating a complete and comprehensive theory of development is like packing a suitcase for a trip: The clothes are too many, the suitcase is too small, and something will remain out. Moreover, a theory of development faces several problematic aspects [122,125,126].

The first problem is a definitory one. Every definition of development leaves something out [127,128]: Some processes are often overlooked (regeneration); some definitions are biased (adultocentric perspective); some other definitions are too restricted (excluding unicellular organisms), and so on.

However, we believe that several aspects may be useful in the elaboration of a theory of development, in particular, and of a more comprehensive and complete EvoDevo Synthesis, in general.

The first aspect concerns the boundaries, both temporal and spatial, of development. From a spatial point of view, development involves multiple resources distributed on multiple levels of organization, as we have seen in the various accounts of EvoDevo. From a temporal point of view, development does not end with entering the adult stage [129,130]. Furthermore, development can go further than the boundaries of generations involving the influences (physiological, behavioral, etc.) of one generation over another (parental effects) [130]. Development is also a scaffolded process, in which several factors act as informational resources in the developmental processes (holobiont scaffolding) [131] (also cultural and cognitive scaffolding) [132].

A second aspect is represented by a broader understanding of the processes and mechanisms affecting the genetic level [133]. The discovery of Hox genes and of the genetic regulatory systems represented a milestone for EvoDevo, and the importance of such mechanisms in development is beyond question. However, it is one thing to acknowledge the importance of a discovery and of a mechanistic explanation, it is another to attribute exclusive causal powers to such mechanisms; especially in the light of a post-genomic conception of the gene [134]. The gene has become a complex entity [135], and the genome, far from being a simple repository of developmental and evolutionary information, is conceived as a Read-Write data storage system [136]. So, genes and genomes must be conceived of in a more dynamic and reactive way, not as exclusive protagonists of development, but as main players always in interaction with their context [137,138].

It has been observed that EvoDevo "[. . .] by itself cannot represent an alternative to the classical paradigm because it has no independent theory of heredity and population dynamics" [139] (p.276). However, it is possible to respond to both observations. As for the absence of an independent theory of inheritance within the EvoDevo, we believe it is a false problem for two reasons. The first reason is quite trivial: Why should there be a need for an independent theory of heredity? The second reason is that the available theories of heredity integrate relatively easily with an appropriate development theory in which different resources (genetics, epigenetics, behavioral, cultural) are involved in development processes [140,141]. Once again, what is needed seems to be greater theoretical integration between the already available perspectives.

As for a theory that takes into account population dynamics, we believe that this is a serious challenge for the EvoDevo. The core issue is that the concept of genes used in population genetics and in MS differs substantially from the concept of genes used in Developmental Synthesis [142]. The ontologies to which the two frameworks refer are profoundly different. Lenny Moss distinguishes between Gene-P and Gene-D, which captures well the problematic nature of this situation [143]. Gene-P is a preformationist conception of genes, in which starting from the genetic level, it is possible to predict the phenotypic outcomes—according to the tradition of genetic transmission [143]. While Gene-D is the developmental gene characterized by a surprising complexity both in its molecular structure and in its interactions with other genes (genetic networks). Overcoming the contrast between these ontologies is the main challenge for the future of EvoDevo [18].

In conclusion, we would like to summarize some issues that have emerged during this discussion and suggest some tasks for the future. The typology of theoretical perspectives is a useful starting point that allows us to consider various approaches to the study of phenomena affecting Development and Evolution. One task for the future is to integrate these perspectives into each other so that we can lay the foundations for subsequent theoretical synthesis. This theoretical synthesis, whose ultimate aim is to explain how Development, Evolution, and Inheritance are linked, requires that a theory of development be developed in the light of the variety of phenomena and approaches presented by the various theoretical perspectives. We want to argue that an EvoDevo resulting from a theoretical integration between the various theoretical perspectives we have presented and strongly focused on the formulation of a theory of development may represent a source of surprises for evolutionary theory.

Funding: This research received no external funding.

Acknowledgments: I would like to thank Alessandro Minelli for inviting me to contribute to the Special Issue "Renegotiating Disciplinary Fields in the Life Sciences" and for his suggestions on a draft version of this paper.

I would also like to thank Daphne Liao and Teodora Todorovic for their editorial support. Finally, I would like to thank the three anonymous reviewers for their useful criticism and suggestions.

Conflicts of Interest: The author declares no conflict of interest.

References

1. Toulmin, S. *Human Understanding, Vol. I, General Introduction and Part I*; Clarendon Press: Oxford, UK, 1972; ISBN 9780691071855.
2. Darwin, C. *The Origin of Species by Means of Natural Selection, or the Preservation of Favored Race in the Struggle of Life*; Cambridge University Press: New York, NY, USA, 2009; ISBN 9781108005487.
3. Robert, S.J. *Embryology, Epigenesis, and Evolution: Taking Development Seriously*; Cambridge University Press: Cambridge, UK, 2004; ISBN 9780521030861.
4. Kirschner, M.W.; Gerhart, J.C. Facilitated Variation. In *Evolution the Extended Synthesis*; Müller, G.B., Pigliucci, M., Eds.; The MIT Press: Cambridge, MA, USA, 2010; pp. 253–280. ISBN 9780262315142.
5. Dobzhansky, T. *Genetics and the Origin of Species*; Columbia University Press: New York, NY, USA, 1951; ISBN 1120819306.
6. Dobzhansky, T.; Ayala, J.F.; Stebbins, G.L.; Valentine, J.W. *Evolución*; Edicion Omega: Barcelona, Spain, 1983.
7. Withfield, J. Postmodern Evolution? *Nature* **2008**, *455*, 281–284. [CrossRef]
8. Hall, B.K. Unlocking the Black Box between Genotype and Phenotype: Cell condensation as Morphogenetic (modular) Units. *Biol. Philos.* **2003**, *18*, 219–247. [CrossRef]
9. Waddington, C.H. *The Evolution of an Evolutionist*; Edinburgh University Press: Edinburgh, UK, 1975.
10. Hall, B.K. Evo-devo or devo-evo—Does it matter? *Evol. Dev.* **2000**, *2*, 177–178. [CrossRef] [PubMed]
11. Gilbert, S.F. Evo-Devo, Devo-Evo, and Devgen-Popgen. *Biol. Philos.* **2003**, *18*, 347–352. [CrossRef]
12. Müller, G.B. Evo-Devo: Extending the evolutionary synthesis. *Nat. Rev. Genet.* **2007**, *8*, 943–949. [CrossRef]
13. Müller, G.B.; Wagner, G.P. Novelty in Evolution: Restructuring the concept. *Annu. Rev. Ecol. Syst.* **1991**, *22*, 229–256. [CrossRef]
14. Müller, G.B.; Newman, S.A. The innovation triad: An EvoDevo agenda. *J. Exp. Zool.* **2005**, *304*, 487–503. [CrossRef]
15. Müller, G.B. Homology: The evolution of morphological organization. In *Origination of Organismal Form: Beyond the Gene in Developmental and Evolutionary Biology*; Müller, G.B., Newman, S.A., Eds.; MIT Press: Cambridge, MA, USA, 2003; pp. 51–69.
16. Minelli, A.; Fusco, G. Homology. In *The Philosophy of Biology: A Companion for Educators*; Kampourakis, K., Ed.; Springer: Dordrecht, The Netherlands, 2013; pp. 289–322. ISBN 9789400765375.
17. Carroll, S.B. *Endless Forms Most Beautiful: The New Science of EvoDEvo and the Making of the Animal Kingdom*; Norton: New York, NY, USA, 2005; ISBN 9780393327793.
18. Wagner, G. *Homology, Genes, and Evolutionary Innovation*; Princeton University Press: Princeton, NJ, USA, 2014; ISBN 9780691180670.
19. Kirschner, M.; Gerhart, J. Evolvability. *Proc. Natl. Acad. Sci. USA* **1998**, *95*, 8420–8427. [CrossRef]
20. Minelli, A. Evolvability and Its Evolvability. In *Challenging the Modern Synthesis: Adapatation, Development, and Inheritance*; Huneman, P., Walsh, D., Eds.; Oxford University Press: Oxford, UK, 2017; pp. 211–238. ISBN 978-0199377176.
21. Arthur, W. The interaction between developmental bias and natural selection: From centipede segments to a general hypothesis. *Heredity* **2002**, *89*, 239–246. [CrossRef] [PubMed]
22. Kuratani, S. Modularity, comparative embryology and evo-devo: Developmental dissection of evolving body plans. *Dev. Biol.* **2009**, *332*, 61–69. [CrossRef]
23. Robert, J.S. Evo-Devo. In *The Oxford Handbook of Philosophy of Biology*; Ruse, M., Ed.; Oxford University Press: Oxford, UK, 2014; pp. 291–309. ISBN 9780199737260.
24. Laubichler, M.D. Evolutionary Developmental Biology. In *The Cambridge Companion to the Philosophy of Biology*; Hull, D., Ruse, M., Eds.; Cambridge University Press: Cambirdge, UK, 2008; pp. 342–360. ISBN 9781139001588.
25. Love, A.C. Evolutionary Developmental Biology: Philosophical Issues. In *Handbook of Evolutionary Thinking in the Sciences*; Heams, T., Ed.; Springer: Berlin/Heidelberg, Germany, 2015; pp. 265–283.

26. Müller, G.B. Evolutionary Developmental Biology. In *Handbook of Evolution, Vol. 2: The Evolution of Living Systems (Including Hominids)*; Wuketis, F.M., Ayala, F.J., Eds.; Wiley-VCH: Hoboken, NJ, USA, 2005; pp. 87–115. ISBN 9783527308385.

27. Minelli, A. Evolutionary Developmental Biology does not Offer a Significant Challenge to the Neo-Darwinian Paradigm. In *Contemporary Debates in Philosophy of Biology*; Ayala, F.J., Arp, R., Eds.; Wiley-Blackwell: Hoboken, NJ, USA, 2010; pp. 213–226. ISBN 9781405159982.

28. Laubichler, M.D. Evolutionary Developmental Biology Offers a Significant Challenge to the Neo-Darwinian Paradigm. In *Contemporary Debates in Philosophy of Biology*; Ayala, F.J., Arp, R., Eds.; Wiley-Blackwell: Hoboken, NJ, USA, 2010; pp. 198–212. ISBN 9781405159982.

29. Carroll, R.L. Towards a new evolutionary synthesis. *Tree* **2000**, *15*, 27–32. [CrossRef] [PubMed]

30. Raff, R.A. Evo-Devo: The evolution of a new discipline. *Nat. Rev. Gen.* **2000**, *1*, 74–79. [CrossRef]

31. Minelli, A. Grand challenges in evolutionary developmental biology. *Front. Ecol. Evol.* **2015**, *2*, 1–11. [CrossRef]

32. Müller, G.B. Evo-devo as a discipline. In *Evolving Pathways: Key themes in Evolutionary Developmental Biology*; Minelli, A., Fusco, G., Eds.; Cambridge University Press: Cambridge, UK, 2008; pp. 5–30. ISBN 9781849722698.

33. Hall, B.K.; Olson, W.M. *Keywords and Concepts in Evolutionary Developmental Biology*; Harvard University Press: Cambridge, MA, USA, 2003; ISBN 9781849722698.

34. Müller, G.B. Six Memos for Evo-Devo. In *From Embryology to EvoDevo: A History of Developmental Evolution*; Laubichler, M.D., Maienschein, J., Eds.; The MIT Press: Cambridge, UK, 2007; pp. 499–524.

35. Minelli, A. *Fusco, Evolving Pathways: Key themes in Evolutionary Developmental Biology*; Cambridge University Press: Cambridge, UK, 2008; ISBN 9781107405455.

36. Burian, R.M. On conflicts between genetic and developmental viewpoints—And their attempted resolution in molecular biology. In *Structures and Norms in Science*; Dalla Chiara, M.L., Doets, K., Mundici, D., van Benthem, J., Eds.; Kluwer Academic Publishers: Dordrecht, The Netherlands, 1997; pp. 243–264. ISBN 978-0-7923-4384-4.

37. Laubichler, M.D.; Wagner, G.P. 2004—Introduction to the papers of the 2001 Kowalevsky Medal Winner Symposium. *J. Exp. Zool. Mol. Dev. Evol.* **2004**, *302*, 1–4. [CrossRef]

38. McGinnis, W.; Levine, M.S.; Hafen, E.; Kuroiwa, A.; Gehring, W.J. A Conserved DNA sequence in homeotic genes of the Drosophila Antennapedia and bithorax complexes. *Nature* **1984**, *308*, 428–433. [CrossRef]

39. Wakimoto, B.T.; Kaufman, T.C. Analysis of Larval Segmentation in Lethal Genotypes Associated with the Antennapedia Gene Complex in Drosophila Melanogaster. *Dev. Biol.* **1981**, *81*, 51–64. [CrossRef]

40. Carroll, S.B. Endless Forms: The Evolution of Gene Regulation and Morphological Diversity. *Cell* **2000**, *101*, 577–580. [CrossRef] [PubMed]

41. Slack, J. A Rosetta Stone for pattern formation in animals? *Nature* **1984**, *310*, 364–365. [CrossRef]

42. Gould, S.J. *The Structure of Evolutionary Theory*; The Belknap Press of Harvard University Press: Cambridge, MA, USA, 2002; ISBN 9780674006133.

43. Davidson, E.H. *The Regulatory Genome: Gene Regulatory Networks in Development and Evolution*; Academic Press: New York, NY, USA, 2006; ISBN 9781280641329.

44. Carroll, S.B. Evo-Devo and an Expanding Evolutionary Synthesis: A Genetic Theory of Morphological Evolution. *Cell* **2008**, *134*, 25–36. [CrossRef]

45. Carroll, S.B. 2001 The Big Picture. *Nature* **2001**, *409*, 669. [CrossRef] [PubMed]

46. Sterelny, K. Development, Evolution, and Adaptation. *Philos. Sci.* **2000**, *67*, 369–387. [CrossRef]

47. Futuyma, D.J. Evolutionary biology today and the call for an extended synthesis. *Interface Focus* **2017**, *7*, 20160145. [CrossRef]

48. Hoekstra, H.E.; Coyne, J.A. The Locus of Evolution: EvoDevo and the genetics of adaptations. *Evolution* **2007**, *61*, 995–1016. [CrossRef] [PubMed]

49. Nikla, K.J. Thinking outside the Hox. *Biol. Theory* **2006**, *1*, 128–129. [CrossRef]

50. Medina, M.L. Two "EvoDevos". *Biol. Theory* **2010**, *5*, 7–11. [CrossRef]

51. Amundson, R. *The Changing Role of the Embryo in Evolutionary Thought*; Cambridge University Press: Cambridge, UK, 2005; ISBN 9780521703970.

52. Gehring, W.J.; Hiromi, Y. Homeotic Genes and the Homeobox. *Annu. Rev. Genet.* **1986**, *20*, 147–173. [CrossRef]

53. Gehring, W.J. *Master Control Genes in Development and Evolution: The Homeobox Story*; Yale University Press: New Haven, CT, USA, 1998; ISBN 9780300074093.

54. Forgacs, G.; Newman, S.A. *Biological Physics of the Developing Embryo*; Cambridge University Press: Cambridge, UK, 2005; ISBN 9780521783378.

55. Kauffman, S. *The Origins of Order: Self-Organizating and Selection in Evolution*; Oxford University Press: Oxford, UK, 1993; ISBN 9780195058116.

56. Newman, S.A.; Muller, G.B. Epigenetic mechanisms of character origination. *J. Exp. Zool. B Mol. Dev. Evol.* **2000**, *288*, 304–317. [CrossRef]

57. Callebaut, W.; Muller, G.B.; Newman, S.A. The Organismic Systems Approach: Evo-Devo and the Streamlining of the Naturalistic Agenda. In *Integrating Evolution and Development. From Theory to Practice*; Samson, R., Brandon, R., Eds.; The MIT Press: Cambridge, MA, USA, 2007; ISBN 9780262195607.

58. Müller, G.B.; Newman, S. Origination of Organismal Form: The Forgotten Cause in Evolutionary Theory. In *Origination of Organismal Form: Beyond the Gene in Developmental and Evolutionary Biology*; Müller, G.B., Newman, S., Eds.; MIT Press: Cambridge, MA, USA, 2003; pp. 3–10. ISBN 9780262280327.

59. Müller, G.B.; Newman, S. *Generation, Integration, Autonomy: Three Steps in the Evolution of Homology. Homology (Novartis Foundation Symposium 222)*; Wiley: Chichester, UK, 1999; pp. 65–79. ISBN 9780470515662.

60. Goodwin, B. Evolution and the generative order. In *Theoretical Biology: Epigenetic and Evolutionary Order from Complex Systems*; Goodwin, B., Saunders, P.T., Eds.; Edinburgh University Press: Edinburgh, UK, 1989; pp. 89–100. ISBN 9780801845192.

61. Webster, G.; Goodwin, B. *Form and Transformation: Generative and Relational Principles in Biology*; Cambridge University Press: Cambridge, UK, 1996; ISBN 9780521207430.

62. Ho, M.W.; Saunders, P.T. Beyond Neo-darwinism: And epigenetic approach to evolution. *J. Theor. Biol.* **1979**, *78*, 573–591. [CrossRef]

63. Callebaut, W. Self-organization and optimization: Conflicting or complementary approaches? In *Evolutionary Systems*; Van de Vijver, G., Salthe, S.N., Delpos, M., Eds.; Kluwer: Dordrecht, The Netherlands, 1998; pp. 79–100.

64. Weber, B.H.; Depew, P.J. Natural selection and self-organization. *Biol. Philos.* **1996**, *11*, 33–65. [CrossRef]

65. Fontana, W. Modelling EvoDevo with RNA. *BioEssays* **2002**, *24*, 1164–1177. [CrossRef]

66. Fontana, W. The topology of the possible. In *Understanding Change: Models, Methodologies and Metaphors*; Wimmer, A., Kössler, R., Eds.; Palgrave Macmillan: London, UK, 2006; pp. 67–84. ISBN 9780230524644.

67. Lewontin, R.C. *The Docrtine of DNA: Biology as Ideology*; Penguin: New York, NY, USA, 1993; ISBN 9780140232196.

68. Rose, S.; Lewontin, R.C.; Konin, L. *Not in Our Genes: Biology, Ideology and Human Nature*; Penguin Books: Harmondsworth, UK, 1990; ISBN 9780394508177.

69. Lewontin, R.C. *The Triple Helix: Gene, Organism, and Environment*; Harvard University Press: Cambridge, MA, USA, 2000; ISBN 9780674006775.

70. Levins, R.; Lewontin, R.C. A program for biology. *Biol. Theory* **2006**, *1*, 333–335. [CrossRef]

71. Kohl, P.; Crampin, E.J.; Quinn, T.A.; Noble, D. Systems Biology: An Approach. *Clin. Pharmacol. Ther.* **2010**, *88*, 25–33. [CrossRef]

72. Kitano, H. Systems Biology: Toward System-Level Understanding of Biological Systems. In *Foundations of Systems Biology*; Kitano, H., Ed.; The MIT Press: Cambridge, MA, USA, 2001; pp. 1–36. ISBN 9780262277204.

73. Kirschner, M.W. The Meaning of Systems Biology. *Cell* **2005**, *121*, 503–504. [CrossRef]

74. Krohs, U.; Callebaut, W. Data without models merging with models without data. In *Systems Biology: Philosophical Foundations*; Boogerd, F.C., Bruggeman, F.J., Hofmeyer, S.J.-H., Westerhoff, H.V., Eds.; Elsevier: Amsterdam, The Netherlands, 2007; pp. 181–213. ISBN 9780444520852.

75. Noble, D. The Aims of Systems Biology: Between Molecules and Organisms. *Pharmacopsychiatry* **2011**, *44* (Suppl. 1), 9–14. [CrossRef]

76. Kitano, H. Systems Biology: A Brief Overview. *Science* **2002**, *295*. [CrossRef]

77. Newman, S.A. *The Fall and Rise of Systems Biology: Recovering from a Halfcentury Gene Hing*; GeneWatch: London, UK, 2003; Volume 16, pp. 8–12.

78. Noble, D. A theory of biological relativity: No privileged level of causation. *Interface Focus* **2012**, *2*, 55–64. [CrossRef]

79. Noble, D. *The Music of Life: Biology beyond the Genome*; Oxford University Press: Oxford, UK, 2006; ISBN 9780199295739.

80. Piaget, J. *Behavior and Evolution*; Pantheon Books: New York, NY, USA, 1978; ISBN 9780394735887.

81. Weiss, P. Cellular Dynamics. *Rev. Mod. Phys.* **1959**, *31*, 11–20. [CrossRef]
82. Griesemer, J.R. Reproduction and the reduction of genetics. In *The Concept of the Gene in Development and Evolution: Historical and Epistemological Perspectives*; Beurton, P.J., Falk, R., Rheinberger, H.J., Eds.; Cambridge University Press: New York, NY, USA, 2000; pp. 240–285. ISBN 9780521060240.
83. Griesemer, J. The units of evolutionary transitions. *Selection* **2000**, *1*, 67–80. [CrossRef]
84. Gottlieb, G. Probabilistic Epigenesis. *Dev. Sci.* **2007**, *10*, 1–11. [CrossRef]
85. Griesemer, J.R. Reproduction and scaffolded developmental processes. In *Towards a Theory of Development*; Minelli, A., Pradeu, T., Eds.; Oxford University Press: Oxford, UK, 2014; pp. 182–202. ISBN 9780199671434.
86. Wimsatt, W.C.; Griesemer, J.R. Reproducing entrenchments to scaffold culture. In *Integrating Evolution and Development: From Theory to Practice*; Sanson, R., Brandon, R.N., Eds.; The MIT Press: Cambridge, MA, USA, 2007; pp. 227–323. ISBN 9780262195607.
87. Oyama, S.; Griffiths, P.E.; Gray, R.D. Introduction: What is developmental Systems Theory? In *Cycles of Contingency: Developmental Systems and Evolution*; Oyama, S., Griffiths, P.E., Gray, R.D., Eds.; The MIT Press: Cambridge, MA, USA, 2001; ISBN 97802626506321-12.
88. Oyama, S. *The Ontogeny of Information: Developmental Systems and Evolution*; Duke University Press: Durham, NC, USA, 2000; ISBN 9780822324317.
89. Robert, J.S.; Hall, B.K.; Olson, W.M. Bridging the gap between developmental systems theory and evolutionary developmental biology. *BioEssays* **2001**, *23*, 954–962. [CrossRef]
90. Caporael, L.R. Evolution, Groups, and Scaffolded Mings. In *Developing Scaffolds in Evolution, Culture, and Cognition*; Caporael, L.R., Griesemer, J.R., Wimsatt, W.C., Eds.; The MIT Press: Cambridge, MA, USA, 2014; pp. 57–76. ISBN 9781461952367.
91. Caporael, L.R. The Evolution of Truly Social Cognition: The Core Configurations Model. *Pers. Soc. Psychol. Rev.* **1997**, *1*, 276–298. [CrossRef]
92. West-Eberhard, M.-J. *Developmental Plasticity and Evolution*; Oxford University Press: Oxford, UK, 2003.
93. West-Eberhard, M.-J. Developmental Plasticity and the Origin of Species Differences. *Proc. Natl. Acad. Sci. USA* **2005**, *102* (Suppl. 1), 6543–6549. [CrossRef]
94. Kupiec, K.J. *The Origin of Individuals*; Worlds Scientific Publishing: Singapore, 2009; ISBN 9789812704993.
95. Kupiec, K.J. *L'Ontophylogenese: Evolution des Especes et Developpement de L'individu*; Editions Quæ: Versailles, France, 2012; ISBN 9782759217861.
96. Gottlieb, G. *Individual Development and Evolution: The Genesis of Novel Behavior*; Oxford University Press: New York, NY, USA, 1992; ISBN 9780415648486.
97. Abouheif, E.; Favé, M.-J.; Ibarraran-Viniegra, A.S.; Lesoway, M.P.; Rafiqi, A.M.; Rajakumar, R. Eco-Evo-Devo: The times has come. In *Ecological Genomics: Ecology and the Evolution of Genes and Genomes*; Laundry, C.R., Aubin-North, N., Eds.; Advances in Experimental Medicine and Biology 781; Springer: Dordrecht, The Netherlands, 2014; pp. 107–125.
98. Gilbert, S.F.; Epel, D. *Ecological Developmental Biology: The Environmental Regulation of Development, Health, and Evolution*; Sinauer Associates: Sunderland, MA, USA, 2015; ISBN 9781605353449.
99. Bateson, P.P.G. *Behaviour, Development and Evolution*; Cambridge University Press: Cambridge, UK, 2017; ISBN 9781783742486.
100. Stotz, K. Extended evolutionary psychology: The importance of transgenerational developmental plasticity. *Front. Psychol.* **2014**, *5*, 1–14. [CrossRef]
101. Wilkins, A.S. *The Evolution of Developmental Pathways*; Sinauer Associates: Sunderland, MA, USA, 2002; ISBN 9780878939169.
102. Griesemer, J. Development, Culture, and the Units of Inheritance. *Philos. Sci.* **2000**, 348–368. [CrossRef]
103. Giere, R.N. *Scientific Perspectivism*; The University of Chicago Press: Chicago, IL, USA, 2006; ISBN 9780226292137.
104. Robert, J.S. How developmental is evolutionary developmental biology? *Biol. Philos.* **2002**, *17*, 591–611. [CrossRef]
105. Maynard-Smith, J. *Shaping Life: Genes, Embryos and Evolution*; Yale University Press: New Haven, CT, USA, 1998; ISBN 9780300080223.
106. Nuno de la Rosa, L. Computing the Extended Synthesis: Mapping the Dynamics and Conceptual Structure of the Evolvability Research Front. *J. Exp. Zool. Mol. Dev. Evol.* **2017**, *328*, 395–396. [CrossRef]
107. Winther, R.G. Evo-devo as a Trading zone. In *Conceptual Change in Biology, Boston Studies in the Philosophy and History of Science*; Love, A.C., Ed.; Springer: Dordrecht, The Netherlands, 2015; pp. 459–482. ISBN 9789401794121.

108. Pigliucci, M. Do we need an extended evolutionary synthesis? *Evolution* **2007**, *61*, 2743–2749. [CrossRef]

109. Pigliucci, M.; Muller, G.B. *Evolution: The Extended Synthesis*; The MIT Press: Cambridge, MA, USA, 2010; ISBN 9780262315142.

110. Hall, B.K. Evolutionary Developmental Biology (Evo-Devo): Past, Present, and Future. *Evol. Educ. Outreach* **2012**, *5*, 184–193. [CrossRef]

111. Laland, K.N.; Uller, T.; Feldman, M.W.; Sterelny, K.; Müller, G.B.; Moczek, A.P.; Jablonka, E.; Odling-Smee, J.; Wray, G.A.; Hoekstra, H.E.; et al. Does evolutionary theory need a rethink? *Nat. Cell Biol.* **2014**, *514*, 161–164. [CrossRef]

112. Wray, G.A.; Hoekstra, H.E.; Futuyma, D.J.; Lensky, R.E.; Mackay, T.F.C.; Schluter, D.; Strassmann, J.E. Does evolutionary theory need a rethink? No, all is well. *Nature* **2014**, *514*, 161–164. [CrossRef]

113. Laland, K.N.; Uller, T.; Feldman, M.W.; Sterelny, K.; Müller, G.B.; Moczek, A.P.; Jablonka, E.; Odling-Smee, F.J. The extended evolutionary synthesis: Its structure, assumptions and predictions. *Proc. R. Soc. B Boil. Sci.* **2015**, *282*, 20151019. [CrossRef]

114. Müller, G.B. EvoDevo Shapes the Extended Synthesis. *Biol. Theory* **2014**, *9*, 119–121. [CrossRef]

115. Futuyma, D.J. Can Modern Evolutionary Theory Explain Macroevolution? In *Macroevolution*; Serrelli, E., Gontier, N., Eds.; Springer: Berlin/Heidelberg, Germany, 2015; pp. 29–85. ISBN 978-3-319-15044-4.

116. Fàbregas-Tejeda, A.; Vergara-Silva, F. The emerging structure of the Extended Evolutionary Synthesis: Where does Evo-Devo fit in? *Theory Biosci.* **2018**, *137*, 169–184. [CrossRef]

117. Sarkar, S. Evolutionary Theory in the 1920s: The Nature of the "Synthesis". *Philos. Sci.* **2004**, *71*, 1215–1226. [CrossRef]

118. Love, A.C. Evolutionary Morphology, Innovation, and the Synthesis of Evolutionary and Developmental Biology. *Biol. Philos.* **2003**, *18*, 309–345. [CrossRef]

119. Griesemer, J. Towards a theory of Extended Developments. In *Perspective on Evolutionary and Developmental Biology*; Fusco, G., Ed.; Padova University Press: Padova, Italy, 2019; pp. 319–334. ISBN 978-88-6938-140-9.

120. Craig, L.R. The So-Called Extended Synthesis and Population Genetics. *Biol. Theory* **2010**, *5*, 117–123. [CrossRef]

121. Minelli, A.; Pradeu, T. (Eds.) *Towards a Theory of Development*; Oxford University Press: Oxford, UK, 2014; ISBN 9780199671434.

122. Laubichler, M.D.; Prohaska, S.J.; Stadler, P.F. Toward a mechanistic explanation of phenotypic evolution: The need for a theory of theory integration. *J. Exp. Zool. Part B Mol. Dev. Evol.* **2018**, *330*, 5–14. [CrossRef] [PubMed]

123. Jenner, R.A. EvoDevo's identity: From model organisms to developmental types. In *Evolving Pathways: Key Themes in Evolutionary Developmental Biology*; Minelli, A., Fusco, G., Eds.; Cambridge University Press: Cambridge, UK, 2009; pp. 100–119.

124. Minelli, A. *Understanding Development*; Cambridge University Press: Cambridge, UK, forthcoming.

125. Pradeu, T.; Laplane, L.; Prévot, K.; Hoquet, T.; Reynaud, V.; Fusco, G.; Minelli, A.; Orgogozo, V.; Vervoort, M. Defining "Development". *Curr. Top. Dev. Biol.* **2016**, *117*, 171–183. [CrossRef]

126. Minelli, A. *The Development of Animal Form*; Cambridge University Press: Cambridge, UK, 2003; ISBN 9780521808514.

127. Minelli, A. *Forms of Becoming: The Evolutionary Biology of Development*; Princeton University Press: Princeton, NJ, USA, 2009; ISBN 9780691135687.

128. Minelli, A. Animal Development, an Open-Ended Segment of Life. *Biol. Theory* **2011**, *6*, 4–15. [CrossRef]

129. Badyaev, A.V.; Uller, T. Parental effects in ecology and evolution: Mechanisms, processes and implications. *Philos. Trans. R. Soc. B Biol. Sci.* **2009**, *364*, 1169–1177. [CrossRef]

130. Chiu, L.; Gilbert, S.F. The Birth of the Holobiont: Multi-species Birthing Through Mutual Scaffolding and Niche Construction. *Biosemiotics* **2015**, *8*, 191–210. [CrossRef]

131. Caporael, L.R.; Griesemer, J.R.; Wimsatt, W.C. *Developing Scaffold in Evolution, Culture and Cognition*; The MIT Press: Cambridge, MA, USA, 2014; ISBN 9781461952367.

132. Griffiths, P. *Stotz, Genetics and Philosophy: An Introduction*; Cambridge University Press: Cambridge, UK, 2013.

133. Griffiths, P.E.; Stotz, K. Genes in the Postgenomic Era. *Theor. Med. Bioeth.* **2006**, *27*, 499–521. [CrossRef]

134. Gerstein, M.B.; Bruce, C.; Rozowsky, J.S.; Zheng, D.; Du, J.; Korbel, J.O.; Emanuelsson, O.; Zhang, Z.D.; Weissman, S.; Snyder, M. What is a gene, post-ENCODE? History and updated definition. *Genome Res.* **2007**, *17*, 669–681. [CrossRef]

135. Shapiro, J.A. How life changes itself: The Read–Write (RW) genome. *Phys. Life Rev.* **2013**, *10*, 287–323. [CrossRef]

136. Gilbert, S.F. The Reactive Genome. In *Origination of Organismal Form: Beyond the Gene in Developmental and Evolutionary Biology*; Muller, G.B., Newman, S.A., Eds.; The MIT Press: Cambridge, MA, USA, 2003; pp. 87–101. ISBN 9780262280327.

137. Lamm, E. The genome as a developmental organ. *J. Physiol.* **2014**, *592*, 2283–2293. [CrossRef] [PubMed]

138. Muller, G.B.; Pigliucci, M. Extended Synthesis: Theory Expansion or Alternative? *Biol. Theory* **2010**, *5*, 275–276. [CrossRef]

139. Jablonka, E.; Lamb, M.J. *Evolution in Four Dimensions: Genetic, Epigenetic, Behavioral, and Symbolic Variation in the History of Life*; The MIT Press: Cambridge, MA, USA, 2005; ISBN 9780262525848 0262525844.

140. Bondurianky, R.; Day, T. *Extended Heredity: A New Understanding of Inheritance and Evolution*; Princeton University Press: Princeton, NJ, USA, 2018; ISBN 9781400890156.

141. Gilbert, S.F. Genes Classical and Genes Developmental: The different use genes in Evolutionary Synthesis. In *The Concept of the Gene in Development and Evolution: Historical and Epistemological Perspectives*; Falk, R., Rheinberger, H.-J., Eds.; Cambridge University Press: New York, NY, USA, 2000; pp. 178–192. ISBN 0521060249 9780521060240.

142. Moss, L. *What Genes Can't Do*; The MIT Press: Cambridge, MA, USA, 2003; ISBN 9780262632973.

143. Falk, R. The Gene: A concept in tension. In *The Concept of the Gene in Development and Evolution: Historical and Epistemological Perspectives*; Falk, R., Rheinberger, H.-J., Eds.; Cambridge University Press: New York, NY, USA, 2000; pp. 178–192. ISBN 0521060249 9780521060240.

Publisher's Note: MDPI stays neutral with regard to jurisdictional claims in published maps and institutional affiliations.

 *philosophies*

Review

EvoDevo: Past and Future of Continuum and Process Plant Morphology

Rolf Rutishauser

Department of Systematic and Evolutionary Botany, University of Zurich, CH-8008 Zurich, Switzerland; rutishau@systbot.uzh.ch

Received: 7 October 2020; Accepted: 26 November 2020; Published: 1 December 2020

Abstract: Plants and animals are both important for studies in evolutionary developmental biology (EvoDevo). Plant morphology as a valuable discipline of EvoDevo is set for a paradigm shift. Process thinking and the continuum approach in plant morphology allow us to perceive and interpret growing plants as combinations of developmental processes rather than as assemblages of structural units ("organs") such as roots, stems, leaves, and flowers. These dynamic philosophical perspectives were already favored by botanists and philosophers such as Agnes Arber (1879–1960) and Rolf Sattler (*1936). The acceptance of growing plants as dynamic continua inspires EvoDevo scientists such as developmental geneticists and evolutionary biologists to move towards a more holistic understanding of plants in time and space. This review will appeal to many young scientists in the plant development research fields. It covers a wide range of relevant publications from the past to present.

Keywords: process philosophy; scientific perspectivism; developmental genetics; plant structure ontology; homology; land plant phylogeny; morphological misfits; flower; phyllotaxis; *Utricularia*

1. Introduction

Plant morphology is the study of the physical form and structure of plants. Literally, the term "morphology" is derived from the Greek roots: *morphe*, which means form and/or structure, and *logos*, meaning discourse or investigation [1]. Plant morphology usually includes plant anatomy, which is the study of the internal structure of plants, as seen at the microscopic level. Plant morphology comprises both adult structures as well as their development. Two aspects of plant development can be distinguished: (i) the development of whole plants from embryos to fertile adults, known as ontogeny; and (ii) the development of each part, such as a foliage leaf or a flower that starts as a group of undifferentiated stem cells (called meristems or primordia). Unlike most multicellular animals, plants usually have an open bauplan with long-lasting apical growth, continued branching, and repetition of subunits (plant organs) such as leaves, stems, roots, and flowers. This is especially true for seed plants, whereas ferns (as another group of vascular plants) lack flowers and seeds (see Section 4.1).

This essay is written by a plant morphologist and historian of botany. It is a contribution to a special issue in the journal PHILOSOPHIES, edited by Alessandro Minelli on "Renegotiating Disciplinary Fields in the Life Sciences", with another three contributions [2–4]. The four PHILOSOPHIES papers demonstrate the need for scientific pluralism in biological disciplines such as evolutionary biology, developmental genetics, and taxonomy. They focus on recent and ongoing debates in biology (including philosophy of biology) that reveal shortcomings of traditional explanatory frameworks. The gene-centered perspective should be complemented by an organism-centered biology based on developmental processes. My contribution—focussing on plants—emphasizes that our knowledge of comparative biology, especially comparative plant morphology, is based on two centuries of knowledge and should not be neglected by the new generation of EvoDevo scientists.

The second part of this essay is a general introduction on process philosophy and continuum views in biology. It prepares the way of thinking (including scientific perspectivism) needed for a paradigm shift in plant morphology as a valuable sub-discipline of EvoDevo.

The third part is dedicated to Agnes Arber and Rolf Sattler as early proponents of continuum and process morphology in plants. Process thinking and continuum approaches have implications for molecular and developmental genetics. Instead of asking which genes control the development of each single plant part (such as a foliage leaf or a flower), we may concentrate on combinations of developmental processes that create the whole plant body with, for example, leaves and flowers as iterated products.

In the fourth part, various case studies from the plant kingdom will be presented illustrating the need for process thinking and continuum approaches. This part focusses on several new publications on vascular plants elucidating the regulation of plant architecture by genes. For example, the bladderworts (genus *Utricularia*) are a well-known group of carnivorous flowering plants that attract students because of their unique features in biology, morphology, and developmental genetics. More specifically, the aquatic *Utricularia gibba* and the terrestrial *U. reniformis* were chosen as model organisms by geneticists, allowing further insights into developmental and evolutionary processes that may be relevant for land plants in general.

2. Philosophy of Biology: Complementarity, Continuum, and Process Thinking in Development and Evolution

2.1. Philosophy of Biology

Is there any connection between philosophy *and* biology? This question is asked by Rolf Sattler (*1936), who has taught philosophy of biology for many years at McGill University (Montreal) and published a book on that subject [5]. Biological research is influenced by philosophical assumptions such as either/or logic, fuzzy logic, structure/process dualism, or its transcendence. Thus, there are interactions between philosophy and empirical findings. These interactions are the subject of what has been referred to as biophilosophy [5,6].

Sattler [7] writes: "It seems impossible to carry out any scientific investigation without philosophical assumptions. These assumptions are often taken for granted, and scientists may not even be aware of them. Therefore, it is one of the tasks of philosophy of science and biophilosophical investigations to make us aware of these assumptions that usually are part of a worldview. Unless one is aware of these assumptions and the associated worldview, one cannot evaluate them and perhaps change them if they appear inappropriate and no longer supported by empirical evidence."

Agnes Arber (1879–1960) was a British biophilosopher, historian of botany, and plant morphologist [8–10]. At the 16th International Botanical Congress in 1999, a symposium was held on the relationship between Arber's work and new explanatory models for plant development. The symposium contributions were published in *Annals of Botany* (see e.g., [11–15]). Bruce Kirchoff [11] (p. 1103), one of the organizers of the symposium, noted that new techniques in biology "have shifted our research focus toward molecular, physiological, and genetic aspects of plant development. What is forgotten in times like these is that our interpretation of the data is contingent on the models we bring to them. Data alone do not support or refute any theory. Agnes Arber's work is a powerful reminder of this fact. She constantly reminds of the relationship between data and theory by returning again and again to the data to view them in new ways, reinterpreting them in the light of new theories." The systems biologist Arthur Lander [16] stated more recently: "Models do not arise by logical inference from data; they are acts of human creation."

2.2. Scientific Perspectivism and Complementarity in Biology

"Individual scientists, and science as a whole in its specific historical context, only ever see a small proportion of all perceivable events or phenomena. Our understanding always remains incomplete and biased by our personal history and societal context." Johannes Jaeger (2017, p. 139) [17]

Various philosophers and scientists accepted two or more complementary views, perspectives, or modes to describe and interpret form and function of living matter, including growth of plant structures [2,5,17–22].

Scientific perspectivism accepts different positions as perspectives on reality that complement one another. Although perspectivism is often used in colloquial speech, it is not common in natural science including biology [4,5,13,23]. Perspectivism accepts every insight into nature as one perspective (but not the only one) to see and explain biological phenomena. Different perspectives (also called approaches, models) complement each other, rather than compete with each other, although not all of them are meant to be equal approximations to what really occurs in nature. Perspectives may be even somewhat contradictory to each other. As long as they are supported by facts, they should be accepted as explanatory hypotheses, because they illuminate other facets of reality that may be useful and important as well.

For example, Minelli [24] and Pradeu [25], in a historical and evolutionary context, discussed the question: What is an individual in biology? Minelli's conclusion: "None of the concepts thus far advanced by biologists or philosophers of life covers in satisfactory way all instances and aspects of biological individuality". Thus, two or more complementary views may be justified for description of modular organisms, such as a coral or a branched woody plant: Each one may be viewed as a population of many individuals as well as one but highly complex individual [26,27].

Perspectivism was anticipated by Arber [28–30] and earlier philosophers who stressed the "coincidence of contraries". This term describes the somewhat astonishing situation that even biologists are allowed to label phenomena of living organisms (e.g., a leaflet of a plant) with seemingly contradictory terms. Perspectivism is closely related to as-if-ism, pluralism, conceptual nominalism, and Woodger's [31] map analogy.

Examples of complementary perspectives or world views as used in plant morphology will be presented in Section 3.3.

2.3. Clear-Cut Language Versus Fuzzy Concepts in Natural Sciences Especially Biology

In contrast to perspectivism, there is the view based on conceptual realism (essentialism), assuming that structural categories are immanent in life, forming crisp or distinct sets, i.e., terms with non-overlapping connotations and without intermediates. This school of thought (also called substance ontology or substance metaphysics) is common in biology and other natural sciences. Looking for essential "units" coincides with "either/or thinking" as expressed by Aristotle's Law of Contradiction ("A cannot be both A and not-A"), which is the basis of most ordinary discursive–logical reasoning [20,30,32,33]. According to this approach, natural sciences require a clear-cut language consisting of well-defined terms and notions that allow either/or decisions (as exemplified by the plant structure ontology, see Section 3.1).

No doubt, we need a consistent terminology in many disciplines of natural sciences in order to become understandable, even in plant morphology. However, drastic evolutionary changes in bauplans of living organisms may require fuzzy rather than clear-cut concepts of organ identity for description [4,13,34–37], calling for a fuzzy approach while searching for continua in time and space. This second approach coincides with process philosophy (also called process metaphysics) in cognitive psychology and systems biology. For an illustration, let us take two examples from colloquial speech: (i) The concept of biological individuality [24,25], which applies well to most human beings (except for Siamese twins), becomes fuzzy when used to describe colonial organisms, including corals and trees (and even social insects such as the honeybee). (ii) The behavioral biologist Bernhard Hassenstein (1922–2016) [23] pointed out that life must be seen as "injunction", i.e., as a concept that cannot be defined by a clear-cut set of properties. Illustrating the problem, Hassenstein asked the question "How many grains result in a heap?" There is no clear-cut answer to this question. It depends on our perspective (including the size of the grains), if five, 20, or 50 grains are needed as a minimum to get a heap.

2.4. From Process Philosophy to Process Thinking in Biology

The tradition of substance ontology has dominated philosophers' views of reality so far. However, it seems that times may be changing. According to Sattler [7]: "A structure is not seen as having processes, a structure is seen as process(es)." Then, there is no longer a structure–process dualism. Process philosophy argues that processes are more fundamental than substance (i.e., static things or entities) [38,39]. Jaeger [17] (p. 138) gives a short introduction to process philosophy as part of systems biology: "In its most pragmatic form, process philosophy states that it is useful and important to study nature in processual terms. In other words, while it is important to know what a system is made of, it is the interactions between its components that define a system's behavior (its dynamical repertoire) and its potential for future change . . . Thus, things can only be studied as parts of processes, which is exactly what systems biology is supposed to do."

This process philosophical approach was used by Nicholson and Dupré [40] for all kind of organisms (even for viruses): "The living world is a world of process rather than a world of things." According to the process(ual) philosophy of biology, there are no misfits; everything fits, because everything can be understood in terms of developmental processes such as, for example, branching and articulation. Baedke and Mc Manus [41] are quite aware of this trend towards process thinking in biosciences when they write: "Recently, a growing number of philosophers of science, first and foremost John Dupré and colleagues [40,42,43], have emphasized the need to adopt a processual perspective for studying the complex dynamics and close connectedness of living systems."

Alfred North Whitehead (1861–1947) was one of the founders of modern "process philosophy" [38,39,44,45]. However, when searching for Whitehead's influence on well-known evolutionary biologists such as Ernst Mayr, one is disappointed. Mayr [46] (p. 38) wrote only one sentence about Whitehead, calling him "a peculiar mixture of a mathematician and a mystic" (in German "eine sonderbare Mischung von Mathematiker und Mystiker"). The science historian Ilse Jahn [47] in her superb "Geschichte der Biologie" did not mention either "process philosophy" or Whitehead's contributions at all.

There is a growing group of botanists who internalized process thinking in their research attitudes during the last two to four decades. Mabberley and Hay [48] (p. 122) wrote: " . . . it has become increasingly accepted that in plants, it is better to consider an organism and what are perceived by many as its component parts at any one time as a snapshot in a dynamic transformation of growth, maturity, and senescence." In botany, process thinking is known as process plant morphology or dynamic morphology [49] (see more in Section 3.6).

Johanna Seibt [33] admits: "While process philosophers insist that all within and about reality is continuously going on and coming out, they do not deny that there are temporally stable and reliably recurrent aspects of reality." In biosciences, there is a diversity of hierarchical descriptions with respect to structures (see e.g., plant structure ontology, as mentioned in Section 3.1.). Baedke and Mc Manus [41,50] have shown that developmental processes in organisms can also be viewed as part of time scale hierarchies (labelled as "dynamic hierarchies"), including processual hierarchies between genotype and phenotype. Somewhere between the structural approach (substance ontology) and process philosophy (process ontology) is the concept of dynamical patterning modules, as described by Benitez et al. [51]. These are defined as "sets of conserved gene products and molecular networks in conjunction with the physical morphogenetic and patterning processes they mobilize." Similar "developmental units" were proposed for modular organisms, including land plants. Minelli [4] called them "organizational modules"; Lacroix et al. [49], following the Russian paleobotanist S.V. Meyen, labelled them as "repeating polymorphic sets".

2.5. Evolutionary Developmental Biology (EvoDevo)

"Organization is a continuum in the physical world. Organization is also a continuum in the ontogenesis and reproduction of the individual organism and in the phyletic line of which it is a component." C.W. Wardlaw [52] (p. 371)

Evolution involves the transformation of ontogenies. Sattler [7] argues as follows: "A problem of much evolutionary morphology is that, contrary to what is often said, structures are not directly derived from one another in a strict genealogical sense: organisms give rise to one another, but not structures ... Although this seems obvious, it has often been ignored, but now it appears to be increasingly recognized in EvoDevo (evolutionary developmental biology): ontogenies change during evolution. Zimmermann [53] knew this long ago when he emphasized hologeny, which refers to the continuum of successive ontogenies. He saw clearly that evolution involves the transformation of ontogenies."

Within the last 30–40 years, evolutionary biology/morphology became known as EvoDevo by the arrival of developmental and molecular genetics. Minelli [4] writes: "One of the most sensational results was the discovery of the involvement of homologous genes in the development of such different organisms as mouse and fruit fly. The increasingly accessible contents of the black box between genotype and phenotype proved to be of utmost interest not only for development biologists, but also for evolutionary biologists". According to Minelli [4], the emergence of EvoDevo as a new interdisciplinary research field is conventionally fixed by the books of Raff and Kaufman [54] and Hall [55]. A concise introduction to EvoDevo is given by Minelli [56], pointing to the following four aspects [57–61]:

(i)   In the last 20 years since 2000, there has been a rapid growth of EvoDevo as a new approach to understanding the evolution and development of organismic form.

(ii)   To a considerable extent, EvoDevo deals with developmental genes, their evolution, and their expression.

(iii)   EvoDevo explains the arrival of the fittest, whereas Darwinism explains its survival.

(iv)   There is a strong need to focus on the phenotype, which is at the same time the product of development and the direct target of selection.

Ivan Amato [3] in his new PHILOSOPHIES essay tried to answer the following questions: "From an epistemological point of view, what is the relationship between EvoDevo and previous biological tradition? Is EvoDevo the carrier of a new message about how to conceive evolution and development?" It is worth reading Amato's review, because EvoDevo can be defined differently. While referring to new evolutionary theories in and beyond neo-darwinism, Amato presents 17 (seventeen!) different perspectives that complement each other to some degree, among them systems biology [17], epigenetic EvoDevo sensu Müller and Newman [62], and self-organization of biological complexity [63,64].

3. Classical, Continuum, and Process View in Plant Morphology

3.1. Classical Plant Morphology as the Tradional Mainstream View

It was not the very versatile Goethe (being poet as well as scientist), but mainly the German plant morphologist Wilhelm Troll (1897–1978) [65], who accepted [66] (p. 246): "that leaves, shoots, and the like are fundamental building blocks of the plant archetype: the starting point (in the ideal sense rather than the modern temporal/evolutionary sense) from which all plants are derived." Traditional botanists such as Wilhelm Troll were used to neatly distinguish foliage leaves (phyllomes), stems (caulomes), and roots as the three most obvious structural categories of vascular plants, as distinct and invariant modules that build up higher plants. Classical plant morphologists ("typologists") usually take for granted that all (or most) multicellular vascularized structures found in ferns and seed plants can be identified (homologized 1:1) with either a leaf, a stem, or a root [15,67,68].

The developmental geneticists Kinoshita and Tsukaya [69] mentioned the position criterion for the distinction of leaves and stems (shoots) in flowering plants and gymnosperms: "The aerial part of (typical) seed plants is called the shoot, which is composed of stems, leaves, and axillary buds. These are produced by indeterminate activity in the shoot apical meristem (SAM), whereas the morphogenesis of leaves depends on determinate activity of leaf meristems." Based on the publications by Kaplan [67,68], Dengler and Tsukaya [70], and Frangedakis et al. [71], the developmental geneticists Cruz et al. [72] (p. 1) summarized the tenets of classical plant morphology as follows: "Vascular plant organs are classically defined based on their position; on their tissue organization (symmetry axes and vascular tissue); and on the presence, position, and activity of their meristems. With these criteria, leaves are lateral determinate organs generally with an abaxial–adaxial asymmetry, and these features seem to generally apply well to leaves in seed plants. On the other hand, shoots are characterized by indeterminacy and are marked by the expression of *Class I KNOTTED-LIKE HOMEOBOX (KNOX)* genes in the shoot apical meristem (SAM)."—It is interesting to realize that the very last criterion to distinguish stems (shoots) and leaves mentioned by Cruz et al. [72] is provided by developmental genetics (see Section 3.5).

Mainly based on classical plant morphology is the **Plant Structure Ontology (PSO: www. plantontology.org)**, which is thought to be a controlled vocabulary of botanical terms describing morphological and anatomical structures representing organ, tissue, and cell types and their hierarchical relationships [34,73]. It was developed by the Plant Ontology Consortium in response to the rapid proliferation of molecular sequences and databases, which has created data access problems for biologists. The main intent was to create a unified and hierarchical vocabulary of plant morphology and anatomy that can be used to describe spatial and temporal aspects of gene expression, because—traditionally—the same terms are sometimes applied to different plant structures in different taxonomic groups. Each of the nearly 60,000 structural terms is linked to a PO number. For illustration of the complexity of the PSO, let us start with the definition of a **whole plant** (PO: 0000003): "A plant structure (PO:0005679) that is a whole organism." Now, we need to know what a **plant structure** is. PSO provides the following definition (PO:0009011): "An anatomical structure that is or was part of a plant, or was derived from a part of a plant." Let us go on with the definition of **plant organ** (PO:009008) as provided by PSO: "A multi-tissue plant structure (PO:0025496) that is a functional unit, is a proper part of a whole plant (PO:0000003), and includes portions of plant tissue (PO:0009007) of at least two different types that derive from a common developmental path." Then let us continue with **cardinal organ part** (PO:0025001): "A cardinal part of multi-tissue plant structure (PO:0025498) that is a proper part of a plant organ (PO:0009008) and includes portions of plant tissue (PO:0009007) of at least two different types." Now we look for the PSO definitions of **leaf** as one of the three main plant organs in classical plant morphology: **Phyllome** (PO:0006001) and leaf (PO:0025034) are defined as follows: A phyllome is "a lateral plant organ (PO:0009008) produced by a shoot apical meristem (PO:0020148)". More specifically: A leaf is "a phyllome (PO:0006001) that is not associated with a reproductive structure". With respect to vascular plants, the leaf is specified as **vascular leaf** (PO:0009025), i.e., as leaf "having vascular tissue" with the following circumscription: "In angiosperms, commonly thought of as one of the three basic parts of the seed plant body, a structure usually of determinate growth, without secondary thickening, and of superficial origin, often flattened and photosynthetic in part, and in the axil of which is found a bud. Occurs in the sporophytic phase of a plant life cycle." Let us now look for subunits (e.g., leaflet) of leaves. As an example, the short PSO definition of **compound leaf** (PO:0020043) is "a leaf having two or more distinct leaflets that are evident as such from early in development." As a daughter term, the definition of **leaflet** (PO:0020049) is given: "A cardinal organ part (PO:0025001) that is one of the ultimate segments of a compound leaf (PO:0020043)."

In spite of the original aim of the PSO [73] to provide a "controlled vocabulary of botanical terms describing morphological and anatomical structures", the reader of the paragraph above may have been confused by the hierarchical complexity of structural terms used in botany. A critical review on possible shortcomings of PSO was already presented by Kirchoff et al. [34]. Besides the danger of simplification and some weaknesses due to the hierarchical order of all the structural terms, the PSO does not cover alternative shoot concepts (e.g., phytomere) and bauplan oddities such as leaf-shoot intermediates and other kinds of morphological misfits (see below and Section 3.3).

3.2. Morphological Misfits (Bauplan Oddities) as Test for Classical Plant Morphology

Classical plant morphology as understood by Wilhelm Troll was based on essentialism, which is also known as conceptual realism [46]. According to David Baum [66]: "Essentialism implies that parts, a few more or less universal types, are the universal building blocks of vascular plants".

Schneider [74] presented the following criticism of a too dogmatic way of classical plant morphology: "It is important to keep the influence of classical morphology in mind, because it is still very influential especially as a result of the remarkable efforts of Don Kaplan [67,68], who provided access to this knowledge to a mainly English reading audience. However, classical morphology is not well aligned to the concept of Darwinian evolution. It is very important to keep in mind that the meaning of 'primitive' in a typological context is not synonymous with 'ancestral' in an evolutionary context".

Bauplans (also called body plans) are generalizations of our thinking and classifying mind. However, there is no doubt that certain animals and plants evolved structures (organs, appendages) that cannot be sensibly accommodated in traditional descriptions. Some plant groups were referred to as **morphological misfits** by Adrian Bell [75]. He pointed to the fact that morphological misfits are "misfits to a botanical discipline, not misfits for a successful existence". Morphological misfits are also observable in animals [36]. Various morphological misfits emerged as morphological key innovations (perhaps "hopeful monsters") that gave rise to new evolutionary lines of organisms [76–78]. The concept of "morphological misfits" is an eye-catcher that allows labelling of all kinds of morphological deviations in the wild, mainly based on major genetic changes such as homeosis (i.e., ectopic gene expression in a seemingly wrong position) and other kinds of developmental repatterning [36,79,80]. Morphological misfits are found in various aquatic vascular plants such as the duck-weeds (*Lemna* and allies in Araceae) [81] and the river-weeds (Podostemaceae) [22,82–84], but also in terrestrial plants such as the gloxinia family (Gesneriaceae), containing one-leaf plants (members of the genera *Monophyllaea* and *Streptocarpus*) [69,85] and in plant groups containing both aquatic and terrestrial members such as the bladderworts (*Utricularia*, Lentibulariaceae) [22]. The bladderworts (genus *Utricularia*) belong—also with respect to molecular developmental genetics—to the best-known examples of morphological misfits in vascular plants. The case studies presented in Sections 4.1–4.5 will show the need of continuum approaches and process thinking transcending classical plant morphology.

3.3. Scientific Perspectivism and Complementarity in Plant Morphology

"If we once accept the fact that 'stem' and 'leaf' are no more than convenient descriptive terms, which should not be placed in antithesis as if they corresponded to sharply opposed morphological categories, the problem of their delimitation and of their differentiating characters vanishes into thin air." Agnes Arber 1930 [86] (pp. 308–309)

Baum [66] published a stimulating essay on the key question: **What *are* plant parts?** He focuses on the "dynamic nature of development" and claims—at least for multicellular plants such as vascular plants—that "parts are not sharply defined things like the bolts and nuts and gears of a car, but repeating eddies or vortices in developmental flows." Baum, as an enthusiastic proponent of EvoDevo, presented three complementary perspectives to understand the nature of plant parts such as leaves (or what botanists usually accept as "leaves" in vascular plants): (i) parts-as-structures (linking structures to their genetic causes), (ii) parts-as-functions (linking traits to responses to selection), and (iii) parts-as-processes (allowing to better recognize, e.g., "leaves" along a stem as being serially homologous). Baum [66] (p. 254) concluded: "Plant parts can properly be considered structures, functions, and processes".

Four complementary views to look at leafy shoots in vascular plants. According to Howard [87], the plant is a continuous whole. He emphasized this when he described the shoot of vascular plants as a "stem–node–leaf continuum." With regard to the leafy shoots in vascular plants, there are different perspectives that complement one another. During the two centuries since Goethe, the shoot of vascular plants was dismembered conceptually in (at least) four different ways [32,88,89]:

(I) The **classical stem-and-leaf model**: This is the popular model, as favored by most plant biologists, including classical plant morphologists (see Section 3.2). Stems (caulomes) and leaves (phyllomes) are accepted as structural categories that build up the shoot in seed plants and ferns. We may use this model as a "rule of thumb" (Tomlinson in Sattler [90]).

(II) The **fertile-leaf model**: This less known model accepts each leaf with one or more buds in its axils as one developmental unit, as part of one bifurcating meristem giving rise to both the subtending leaf and the axillary bud(s). This model, going back to Warming (1872) [91] and Goethe (1790) [92], explains why in seed plants (less so in ferns), lateral branches or flowers arise in the axil of leaves and may even form common primordia that bifurcate into a leaf and its axillary bud. Unequal bifurcation of such common primordia resulted in axillary flowers lacking their subtending leaves (nearly) completely, as it is typical for inflorescences in Arabidopsis and other members of the Brassicaceae [93]. Lyndon [94] (p. 21), as a proponent of this model wrote: "It is easy to forget that the leaf is not a single but a dual structure—a leaf with a bud in its axil." Shoot buds or flowers arising on leaves (a phenomenon known as epiphylly) provide further evidence for the validity (or at least the heuristic value) of the fertile-leaf model [95,96] (see Section 4.4).

(III) The **leaf-skin model**: In this model as proposed by Saunders [97] the terms leaf and stem are still accepted as structural categories, but the borderline between them is drawn differently. The stem cortex is conceived as being formed by the elongated leaf bases, which cover the stem core like a skin (see Section 4.3).

(IV) Unlike the above-mentioned three complementary models (I–III) that still accept leaf and stem as primary units of the shoot, each leaf or leaf whorl can be seen as a growth unit with the stem zone just below the node of its insertion, leading to the **phytomeric model**. In this shoot model, the stem zone, called **phytomere** (also written "phytomer"), consists of a leaf, its axillary bud, the node, and one internode below the leaf. There are many botanists and developmental biologists who favor this shoot model [26,27,32,93,98]. For example, Kinoshita and Tsukaya [69] wrote: "In the aboveground portion of a typical seed plant, a shoot is composed of the repetition of a unit called a phytomere consisting of a stem and a determinately growing leaf." Rohweder (1963) [99] and more recently Vita et al. [100] described and illustrated obvious phytomeres in growing shoots of herbaceous flowering plants such as Commelinaceae while focusing on leaf-related vascular tissue inside the stems, somewhat similar to a chain of inverted cones sticking into each other. The term phytomere is also applied to Arabidopsis, e.g., by Müller-Xing et al. [93]: "Phytomeres are metameric units that are composed of internode and node (leaf plus axillary meristem)."

Adherents of these four shoot models in vascular plants have often argued over which model is the correct or appropriate one [32]. Such arguments seem futile to a great extent. They are based on an either/or logic, meaning that any view is either correct or not. We can overcome this attitude by the acceptance of perspectivism and complementarity (see Section 2.2). Then, the different models complement one another, because they present different perspectives. Already Goethe [92] accepted different and contradictory perspectives on plant morphology. Among the four perspectives mentioned above, Goethe embraced three in different contexts, namely (I) the stem-and leaf model, (II) the fertile-leaf model, and (IV) the phytomeric model.

3.4. Historical Roots of Continuum Plant Morphology

As indicated above, the classical approach to plant morphology with a typological view of organ categories is no longer sufficient to explain all possible forms in flowering plants and ferns [6,19,49]. Rolf Sattler, as a young and enthusiastic botanist, regularly attended the conferences on plant morphology held in and around Germany. Fifty years ago (i.e., 1971) Sattler [101] gave an oral presentation on "A new shoot model", questioning the traditional view of German-style classical morphology (also called typology). Sattler was already professor of Botany and Biophilosophy at McGill University, Montreal (Canada). Because he had grown up in Germany and did all his studies (incl. PhD) mainly in Germany, he was well aware of classical plant morphology that was in the 1980s still a very strong botanical discipline in German-speaking countries, including Austria and Switzerland (see Section 3.2). Sattler [101,102] proposed that "leaves" and "stems" in vascular plants are no more clearly delimited structural categories; they can intergrade to considerable degree.

A historical survey leading from classical plant morphology (typology) to dynamic morphology (encompassing both continuum and process approaches) was published by Christian Lacroix et al. (2005) [49]. Sattler's school of continuum morphology indeed had its forerunners. For example, the principle of iteration (or identity-in-parallel) involving both shoot and leaf was formulated over 70 years ago by the British plant morphologist Agnes Arber. She was convinced that the principle of repetition is found everywhere in plants [28–30,49,86,103]. Focusing on vascular plants, Arber was puzzled by the fact that some steps of the developmental pathway of a whole shoot (i.e., leafy stem) can be repeated within a compound leaf. According to Arber [28] (p. 125) "a typical leaf is a shoot in which the apex is limited in its power of elongation and in its radiality." Thus, Arber [28,103] proposed the **partial-shoot theory of the leaf** in vascular plants [13]. Compound leaves may repeat in each part what they have already produced as a developing whole. Arber [28,103] described many examples of vascular plants showing repetition during growth and development. The repeated unit can be totally identical to the one already formed (i.e., complete repetition of a developmental pathway), or the structures formed afterwards may repeat the preceding ones only to some degree (i.e., partial repetition of a developmental pathway) while deviating in other features.

Honoring Agnes Arber, the continuum (and process) approach in plant morphology was labelled as **fuzzy Arberian morphology** (abbreviated "FAM") [13]. "Fuzzy" refers to fuzzy logic, "Arberian" to Agnes Arber. This approach is not only supported by many morphological data, but also by evidence from molecular genetics [72,104] (see Section 4.1). Arber herself also had forerunners. Several botanists in the 19th and the first half of the 20th century referred to intermediates between morphological categories and thus implicitly or explicitly recognized a continuum; see historical surveys in [13,88,102]. Unfortunately, they were often forgotten or ignored in mainstream plant morphology. Even Johann Wolfgang von Goethe [92], who designed the "Urpflanze" with its three main organs (roots, stems, leaves), accepted the basic tenet of Agnes Arber's partial-shoot theory of the leaf when he wrote: "When leaves divide, or rather when they advance from their original state to diversity, they are striving toward greater perfection, in the sense that each leaf has the intention of becoming a branch" (quoted by C. J. Engard in Mueller [105], p. 10).

Arber's [28,103] partial-shoot theory of the leaf parallels Minelli's [106,107] **paramorphism concept** for animal bauplans such as arthropods and vertebrates, according to which "animal appendages can be regarded as evolutionary duplicates of the main body axis, originated by a re-expression on a secondary (lateral) axis of the developmental routines responsible for the production of the main body axis... In terms of gross morphology, this is suggested for example by the segmented structure of the appendages in arthropods (and to some extent also in vertebrates), animals in which the main body axis is similarly segmented."

3.5. Homology and Organ Identity in Classical and Continuum Plant Morphology: Acceptance of Partial Homology and Fuzzy Organ Identities

Homology versus convergence of plant structures: Plant morphology is a comparative and evolutionary discipline, making comparisons between structures in plants that seem to be more or less related [108]. In groups of plants that are closely related, it is usually easy to explain similarities of plant structures as expression of shared ancestry and common genetics. Thus, similar structures are taken as being nearly identical ("homologous") to each other. In comparing distantly related groups of plants (e.g., ferns on one hand and flowering plants on the other hand), such a decision becomes more complex. Then similarity in appearance of certain plant structures may have evolved by convergence, usually an expression of independent adaptation to common environmental pressures.

The perspective of the researcher may determine whether certain bauplans and their modules are taken as homologous (synapomorphic) or seen as results of convergent evolution. For example, Harrison and Morrison [60], in an essay on the origin of vascular plant shoots and leaves, provided convincing evidence that the usually **planar foliage leaves evolved more than once** from three-dimensional branching shoot systems. This happened independently at least three times, in (1) lycophytes (clubmosses and allies) with "microphylls", (2) monilophytes (ferns, including horse-tails and whisk ferns) with "fronds", and (3) seed plants (gymnosperms and flowering plants) with "leaves" sensu stricto [109]. Why have leaves evolved multiple times? This question can be answered hypothetically by pointing to the environmental conditions during early phases of vascular plant evolution more than 400 million years ago [110,111]. Thus, Harrison and Morris [60] (p. 10) presented the following "survival of the fittest" hypothesis: "Polyphyletic leaf origins were coupled with declining atmospheric CO_2 levels, declining global temperatures, increasing stomatal and vein densities in leaves, the evolution of extensive rooting systems, and increasing plant competition for space to acquire environmental resources."

What is homology in a comparative biological context? Conceptual realism (essentialism, substance ontology) in biology often resurface in more or less subtle forms as either/or questions. Such questions are often framed in terms of homology. Homology has been called "morphology's central conception" [112]. Originally, it was defined as "essential similarity, which implied 1:1 correspondence" [80,113–115]. With regard to animals, Owen (1843) [116] (p. 379) defined a homologue as "the same organ in different animals under every variety of form and function". So, how can we determine what is the same organ? Traditionally, three criteria were used to decide if two structures in animals or plants are homologous to each other or not [56,108,117–120]. Owen used the criterion of **relative position** with respect to other structures = topological similarity, including correspondence in developmental origin (1st criterion). Homology may also be defined by the 2nd criterion of **special quality** (i.e., similarity in structural details, including anatomy, physiology, organ functions). Besides the first two criteria (position, special quality), the 3rd criterion of **transitional forms** is often used, emphasizing a transformation series of intermediates between putatively homologous structures. With the advent of developmental genetics, the "molecular players", i.e., **gene expression** behind complex structures or organs (phenotype), may be used as a 4th homology criterion: Homologues often do share developmental pathways, which are based on common gene pathways, especially when related groups of organisms are compared [49,72,104,120–125].

There is no doubt that the study of developmental genetics has added a great deal to our understanding of morphological evolution. If we try to accept 1:1 correspondence between morphological structures and gene functions, it may be useful to follow the four-steps recipe offered by Jaramillo and Kramer [124]: "In order to make comparative expression analyses more meaningful, we need to take into account (1) the evolutionary history of the morphological feature in question through reconstructions on robust phylogenies; (2) the evolutionary history of the candidate gene, because gene duplications are very common (…) and these events may make the occurrence of neo- and sub-functionalization more common among plants than other organisms; (3) a clear definition of the morphological character of interest; in the case of flowers—perhaps the best-understood angiosperm structures—it is important to distinguish each whorl of organs as a distinct character and the identity that they can take (sepaloid organ, petaloid organ, stamen, carpel) as a character state; (4) comparative expression data from multiple genes in a genetic network responsible for the character of interest [121]." (For details, see case study on flowers in Section 4.4).

However, the homology concept as used in animals and plants is a very difficult issue. There are five different levels of homology in biological systems, from serial (repetitive) homology within a modular organism, to historical homology (also called synapomorphy) down to underlying (=latent) homology and deep evolutionary homology at the molecular genetic level [62,108,126,127]. Therefore, Brigandt [120,123] accepted homology as something like a "nomadic concept" (see Minelli [4]): "Once it migrated from comparative anatomy into new biological fields, the homology concept changed in accordance with the theoretical aims and interests of these disciplines."

Organ identity and meristem identity: An organ in multicellular animals and plants is a part of a living organism with a certain set of functions besides its positional and constructional characters. In the context of plant developmental genetics, "organ identity" means the developmental fate of an uncommitted primordium. This term is used in zoology (e.g., [128]) and botany as well. Organ identity can be defined by morphological criteria and by their gene expression pattern, including organ identity genes that sculpt, for instance, the structure of angiospermous flowers (see Section 4.4). The organ identity concept is closely related to the concept of homology; both have multiple and sometimes conflicting meanings, as reviewed by [35,108,119,121,122]. Organ identity as a concept is also used outside the floral region. The vegetative body of vascular plants shows primordia that are committed to a certain developmental fate [109,129] (see Section 4.2). Acquisition of organ identity often happens progressively rather than at once [130]. In ferns and aberrant flowering plants (morphological misfits) such as *Utricularia,* the commitment of a primordium to become a leaf or shoot (including stem) can be considerably delayed [13,131] (see Sections 4.1 and 4.3). The term "identity" is also used for meristems such as shoot apical meristems (SAMs) in vascular plants that change their identity during the development of an individual plant from seedling stage to flowering. Müller-Xing et al. [93] wrote with respect to Arabidopsis: "The identity of the SAM undergoes several changes during the plant's lifecycle and so do the generated organs that cause modifications in the shoot structure, which can be described by metameric units named phytomeres." Thus, the term "meristem identity" characterizes the growth phases of a shoot apical meristem (SAM), with vegetative meristem, inflorescence meristem, and floral meristem as three possible "identities" [35,93,132].

Partial (=mixed) homology as useful concept for both plants and animals: Homology also includes partial homology, as proposed for vascular plants by Sattler [19,113,119,133]. The continuum approach in plant morphology is based on the acceptance of partial homology, as reviewed by Sattler [7]: "I proposed a partial homology concept (as a semi-quantitative homology concept) over 50 years ago (in 1966), but it was not well received by many typologists and evolutionary morphologists who, like typologists, insisted that all homology must be total, that is, 1:1 correspondence (or essential similarity). Partial homology leads in vascular plants to a continuum between structural categories (plant organs) such as, leaves, stems, roots, and even multicellular hairs/trichomes." It seems that the concept of partial or mixed homology is more frequently applied to plants than to animals [56,107]. However, Minelli [80] wrote also with respect to animals: "Taking distance from the traditional, all-or-nothing approach according to which two structures are either homologous or nonhomologous, since the 80s of the past century, several authors have been defending the view that all assessments of homology are by necessity partial." Research in EvoDevo has provided further corroboration for this view [114]. Now we have much evidence for partial homologies, even at the molecular level (see Section 3.8).

3.6. Process Plant Morphology and Morphospace in an Evolutionary Context

"I believe the apparent failure of botanists to internalize process morphology arises because this philosophy requires habits of thought and ways of communicating that are not natural for science in general (Seibt [33]). Due to these possibly hard-wired psychological biases, it is very hard to resist the tendency to perceive plants as objects, drawing us to a language of nouns not verbs." David Baum (2019) [66].

According to classical plant morphology ("typology") sensu Wilhelm Troll [65] and Donald Kaplan [67,68], a plant structure is either this or that, but not both. This classical approach is widely used in the study of model plants with typical organs [13,49] (see Section 3.1). Continuum plant morphology allows the acceptance of intermediates and mixed identities between structural categories (see Section 3.5 above). Both classical plant morphology and continuum plant morphology have shortcomings, because both depend on the recognition and definition (crisp or fuzzy) of structural categories. Process morphology (also called dynamic morphology [49]) may be a way of overcoming these shortcomings, because it allows the replacement of structural categories by process combinations ("developmental routines").

Leaf, stem, and root are often—unconsciously—taken for granted as organs in vascular plants. When we realize that these structural categories are, to some extent, arbitrary concepts, each of them encompassing a certain set of developmental processes, then we are prepared to abandon structural concepts and instead refer to combinations of developmental (morphogenetic) processes that depend—to some degree—on gene regulatory networks. This radical view was proposed by members of the morphological school of Rolf Sattler (including Barabé, Jeune, Lacroix, and Rutishauser), as well as Baum and Langdale [1,6,13,19,49,66,119,134–139].

Theoretical and empirical morphospace. The living forms we perceive and conceive of in the realms of multicellular organisms (animals, plants, fungi) "are only a small subset of the possible forms we could imagine" (Minelli [37]). The theoretical morphospace includes all possible process combinations for seed plants, whereas the empirical morphospace contains only those process combinations that are realized in nature [49,140]. Each axis of the morphospace corresponds to a variable that describes some developmental processes of an organism, or its parts. The use of a single morphospace to which gene expression can be annotated is appealing, especially because its use would remove some terminological problems described above.

Process morphology allows seeing whole plants as combinations of developmental processes instead of more or less arbitrarily assigning plant parts to categories such as root, stem, and leaf. Parameters for process morphology with respect to vascular plant development are growth duration, symmetrization (e.g., acquisition of dorsoventral symmetry), positioning, branching, tropism, rhythm, reiteration, and senescence (see [49,141] for lists of the main processes in plant development).

We may ask with respect to morphological misfits such as bladderworts (*Utricularia*) and river-weeds (Podostemaceae) having unusual morphologies [22]: Is the recognition of developmental processes (e.g., branching patterns and growth patterns) more important than proper definition of structural units, i.e., plant organs such as roots, stems, and leaves? According to Sattler [7] "there are no misfits according to process morphology; everything fits because everything can be understood in terms of branching and articulation. By using developmental processes instead of structural categories, morphological misfits such as bladderworts and river-weeds need no longer be forced into one or the other category."

Hypotheses on bauplan evolution in vascular plants: Schneider [74] addressed the evolution of early land plants (ferns mainly), while focusing on the transformation of structures in time [15,142]. This strictly phylogenetic approach is more likely to be compatible with the developmental, process-oriented perspective of fuzzy Arberian morphology than with the typological classification of classical plant morphology [13,15]. The importance of "transformation of forms" versus "fixed typology" can be illustrated with the discussion on the evolution of leaves. The origin of leaves is frequently discussed, and the nonhomology of "leaves" of liverworts, mosses, lycophytes, ferns, and seed plants is widely accepted (see Section 3.5). However, these statements do not provide us with any understanding of the origin of leaves. Process combinations ("developmental routines") are canalized during land plant evolution [15,53,143–145]. Therefore, a process morphological approach may help understanding the evolutionary origin of bauplans in vascular plants. Walter Zimmermann's [53,143] "telome theory" seems to be still useful as an evolutionary hypothesis in the sense of process morphology, because it proposes a sequence of transformative evolutionary processes, especially (i) unequal branching (overtopping), (ii) rearrangement of lateral branches into a single plane (planation), and (iii) infilling of spaces between branches with laminar tissue (webbing/fusion). According to the telome theory, a dichotomously branching shoot system evolved into a foliage leaf by these three transformations [60,142]. Although Zimmermann's telome theory as evolutionary hypothesis has serious limitations, it is still discussed as a heuristic aid by developmental geneticists in search for the regulation of mechanisms of developmental processes (e.g., [144–147]. It is a gradualistic perspective that breaks the evolutionary origin of complex organs such as fern leaves (fronds) into steps (see Section 4.1, also Jill Harrison's Blog 2018: https://thenode.biologists.com/testing-zimmermanns-telome-theory/ research/).

The step-by-step evolution of leaves (fronds) as proposed by Zimmermann coincides with what was called partial or mixed homology (see Section 3.5) because leaves/fronds probably evolved by a sequence of evolutionary transformations. Of course, these hypothetical transformations in the sense of Zimmermann's telome theory need to be understood in terms of the genetic regulation of plant developmental processes [60,144,148]. According to Schneider [74]: "A process-oriented approach, combining phylogenetic and developmental perspectives, instead of a typological-oriented approach, is the most promising approach to identify the developmental pathways and their transformation in the evolution of land plants." Zimmermann [143] (p. 470) was already an early proponent of process morphology when he wrote: "The telome theory not only applies to 'typical' organs that means to such organs as are classified according to classical morphology under the notion of leaf, shoot, axis, stem, root—but all forms of 'Kormophyta' (i.e., vascular plants, including ferns and seed-plants) are to be traced back to the varying combinations and the different degree of manifestation of a few modifying processes."

Process thinking may even allow to define plant structures such as "shoots" differently (if such a definition is needed at all). Plackett et al. [109] (p. 4) argued about the evolution of land plant shoots as follows: "From a developmental perspective, canonical plant shoots (as generally recognized in vascular plants) can be defined as a process, i.e., they develop iteratively from an apex to produce lateral organs. Using this definition, the gametophores of extant mosses and 'leafy' liverworts can also be classified as shoots, possessing an axial body-plan." This growth behavior contrasts with the "thalloid" bauplan of other liverworts (e.g., *Marchantia*), which possess a usually flattened plant body (comprising several cell layers) with apical growth.

Gilbert and Bolker [149] distinguished between process homology (i.e., homology between developmental processes) and structural homology, which represent different hierarchical levels of homology [107]. Jaramillo and Kramer [124] (p. 69) wrote: "Process homology reflects the common inheritance of developmental genetic pathways or modules that can be co-opted to function in diverse situations. For this reason, process homology is dissociable from structural homology. . . . Novel combinations of such genetic modules may be at the heart of many of Sattler's ([113,119]) examples of partial homology or may promote the evolution of novel organ identities."

Several developmental geneticists are aware of a certain degree of fuzziness in plant development. They have used fuzzy concepts such as "leaf-shoot indistinction" [150], "leaf-shoot continuum model" [151,152], and "mixed shoot-leaf identity" [153]) to describe unusual plant structures. These fuzzy concepts are consistent (or nearly so) with fuzzy Arberian morphology, including both continuum and process morphology [72,104]. James [154] summarized as follows: "It is now widely accepted that . . . radiality (characteristic of most shoots) and dorsiventrality (characteristic of leaves) are but extremes of a continuous spectrum. In fact, it is simply the timing of the *KNOX* gene expression!" With respect to developmental similarities between leaves and shoots in vascular plants (including Arabidopsis), Prusinkiewicz and Runions [155] gave credit to Agnes Arber: "This similarity is consistent with the 'partial shoot theory' (Arber [28]), which emphasizes parallels between the growth of shoots and leaves. The strikingly different appearance of spiral phyllotactic patterns and leaves would thus result not from fundamentally different morphogenetic processes, but from different geometries on which they operate: an approximately paraboloid SAM dynamically maintaining its form vs. a flattening leaf that changes its shape and size."

3.7. Mathematical Tools as a Test for Continuum and Process Plant Morphology

In plant morphology, most structures of vascular plants can easily be assigned to pre-established organ categories. However, there are morphological misfits, i.e., intermediate or deviating structures that do not fit those categories associated with a classical approach to morphology (see Section 3.2.). With the mathematical tool of principal component analysis (PCA), Jeune, Lacroix et al. [138,139,156,157] found a way to unify classical and continuum plant morphology as a synthesis through the space of forms. To integrate the diversity of forms in the same general framework, they constructed a theoretical morphospace based on a variety of modalities, where it is possible to calculate the morphological distance between plant organs. Concentrating on shoot, leaf, leaflet, and trichomes as structural categories (but ignoring the root) Jeune, Lacroix et al. tested the hypothesis that classical morphology (typology) and continuum morphology occupy the same theoretical morphospace. They concluded that the relationship between the two approaches remains a question of weighting of criteria. Typical morphological elements (e.g., shoots, leaves, trichomes), atypical structures (e.g., leaf-like shoots, called "phylloclades"), and particular cases representing exotic structures such as the epiphyllous appendages of *Begonia* and "water shoot" and "leaf" of aquatic bladderworts (*Utricularia*) were placed in the morphospace. The more an organ differs from a typical shoot, the further away it will be from the barycentre of shoots. By giving a higher weight to variables used in classical typology, the different organ categories appear to be separate, as expected. If we do not make any particular arbitrary choice in terms of character weighting, as it is the case in the context of continuum morphology, the clear separation between organs is replaced by a continuum. By using an equal weighting of characters, contradictions due to the ponderation of characters are avoided. Thus, Jeune, Lacroix et al. [138,139,156,157] tend to merge classical plant morphology (mainstream morphology, based on a qualitative homology concept implying mutually exclusive categories) and continuum morphology into the more encompassing process morphology = dynamic morphology (see Section 3.6).

3.8. Possible Lack of One-to-One Correspondence Between Structural Categories and Gene Expression in Higher Plants

"We are still far away from understanding how three-dimensional forms are generated by the genetic system." Diethard Tautz (2019) [158]

We live in a gene-centric world. Many biologists accept the view that the development of living organisms (animals, plants, fungi, …) is controlled mainly by genes. Genetic determinism also governs the thinking of many laymen and journalists with respect to heritability of human health and diseases.

However, we have to consider that the genome (DNA) in the cells of living organisms is a vital but not the only driver of developmental processes, as expressed by Gilbert and Bard [159]: "The fertilized egg inherits DNA; it does not inherit 'genes.' Genes and gene products are constructed anew in each cell in the developing embryo by the relationships between DNA, transcription factors, and RNA-splicing factors. Only certain regions of the genome can be genes in different cell types." For similar statements see also [56,158].

We must know what an organ is before we can investigate gene expression in that organ. If structural categories do not provide adequate descriptions of plant structure, perhaps it is possible to define structures based on developmental genetics. This new way to find homology between structural units was already presented as the 4th homology criterion (see Section 3.5). If there is a one-to-one correspondence between structural units (e.g., roots, leaves, and flowers) and the "molecular players behind the characters" [125], it should be possible to identify the structural units by the expression of well-characterized marker genes. To do this, we need to look for, e.g., **organ identity genes** in order to define the structural categories clearly. For example, the *KNOX/ARP* module (as used by Katayama et al. [82,83], Cruz et al. [72,104]) helps with the identification of the leaf as a determinate unit, and the shoot as an indeterminate module in vascular plants. This approach seems to have promise in the cases where control genes for organ identity have been shown to exist, for example, Pax6, which is often accepted as the "master control gene" for eye development in arthropods and vertebrates [57,160].

There are, however, various difficulties when we try to define plant structures by a set of expressed genes. Adopting this approach for all plant organs leads to some bizarre conclusions, because structural homologues in biological systems do not always have the same underlying molecular genetic machinery [62]. Jaramillo and Kramer [124] present arguments against a too strong belief in genetic determinism, especially in combination with homology: "Genetic information is not always a reliable indicator of homology, especially when only one gene is examined. Our particular concern is the conflation of positional evidence for homology with evidence from identity. Many of the genes that have been studied comparatively are floral organ identity homologs, and it appears that such identity programs can be shifted spatially, resulting in homeosis." Examples from both botany and zoology show that homologous structures can result from different genetic controls [62,107,127,161,162]. For instance, leaflet formation in pinnately compound leaves of many eudicots is correlated with *KNOX1* expression in leaf primordia [163,164]. In contrast, in legumes (Fabaceae) such as pea (*Pisum sativum*), the formation of compound leaves depends on the expression of the PEAFLO gene, the pea homologue of LEAFY from Arabidopsis [152,165].

The lack of correspondence may be due to imprecise morphological homology assessments, but it may also arise from the reuse of existing genetic resources in novel contexts because transcription and signaling factors are often used multiple times in context-specific combinations within an organism [34,166]. Vergara-Silva [137] (p. 260) concluded similarly: "Distinct groups of genes that in principle act in one categorical structure are also expressed in another, and the consequence that this overlapping pattern has on cell differentiation is an effective blurring of the phenotypic boundary between the structures themselves." Developmental genetics underlying morphological continua in flowers, and interesting cases of flower–inflorescence continua will be discussed in Section 4.4 (see [132]).

3.9. The Explanatory Power of Process Morphology in Multicellular Plants

Various plant biologists seem to be aware of the explanatory power of process morphology. Much architectural diversity in vascular plants results from the varied growth patterns of apical and axillary meristems [167]. Computational biologists and mathematicians such as Prusinkiewicz and colleagues [155,168] provided further evidence for the heuristic value of continuum and process plant morphology. As mentioned by Prusinkiewicz and Runions [155]: "The form of plants is not coded directly in their DNA, but is produced by a hierarchy of developmental processes that link molecular-level phenomena to macroscopic forms. Many of these processes have a self-organizing character, which means that forms and patterns emerge from interactions between components of the whole system." For example, a suite of developmental processes leads from the leaf primordium to the mature leaf. Tsukaya and his collaborators [85,169,170] subdivided leaf morphogenesis of Arabidopsis, Antirrhinum, and other angiosperms into subsequent processes (phases). They admitted that the mechanisms regulating each process of morphogenesis, such as leaf determination, establishment of dorsoventrality (e.g., adaxialisation), and polarity recognition, are still badly known. Molecular genetics allows to approach to such processes and relevant developmental pathways that are influenced but not completely controlled by gene networks. Molecular genetic studies in recent decades have considerably expanded our understanding of the mechanism of leaf development (including adaxial–abaxial patterning and lamina flattening) in angiosperms [85,169,170]. Combinations of developmental processes even allow to transcend the leaf–stem (or leaf–shoot) boundary. For example,

Nakayama et al. [171] analyzed in *Asparagus* spp. the developmental genetics of the photosynthetic needle-like organs (labelled as "cladodes") that are—according to classical plant morphology—modified shoots (stems). They observed in developing *Asparagus* cladodes the expression of both SAM-related and leaf-lamina-related genes [85]. This result coincides with comparative developmental studies by Cooney-Sovetts and Sattler [172], who interpreted the leaf-like shoots ("cladodes", "phylloclades") in *Asparagus* and allied genera (*Ruscus, Semele*) in the monocot family Asparagaceae as homeotic mosaics, combining developmental processes of both leaves and shoots (stems).

4. Case Studies: Developmental Genetic Studies Supporting the Continuum View in Plant Architecture

The case studies mainly focus on morphological misfits that do not obey the bauplan rules of classical plant morphology (as introduced above, especially Section 3.2). The case studies point to plant structures that are difficult to explain by a simple one-to-one correspondence between structure and gene function. The genetic study of these organisms will likely reveal that some of the phenotypic fuzziness results from overlapping developmental programs (compare Section 3.5, Section 3.6, Section 3.7, and Section 3.8).

4.1. Ferns: Continuum Between Compound Leaves (Fronds) and Shoots

Many botanists accept the hypothesis that during land plant evolution, megaphylls (including fronds and prefronds) of ferns were derived from branched axial organs ("telome trusses") by flattening ("planation"), webbing ("fusion"), and a switch to determinate growth (see Section 3.6). Prefronds, as found in early vascular plants such as the extinct progymnosperms (e.g., *Archaeopteris*), are three-dimensional branched structures and are seen as phylogenetic precursors of compound leaves, especially fronds of modern ferns [15,53,143,167,173].

Comparative morphologists also studied the developmental patterns of modern-day ferns. Leaves of ferns are often labelled as "fronds", because they have an evolutionary history different from the foliage leaves of seed plants [109,174] (see Section 3.5). Developmental mosaics between leaves and shoots are described for various recent ferns, e.g., *Gonocormus* (*Trichomanes* group) with shoot buds developing from the rachis [175]. To label a group of embryonic cells derived from the shoot apical meristem (SAM) as "leaf" is to label it according to its presumptive developmental fate (organ identity) under undisturbed conditions. However, surgical or chemical treatment may change the developmental fate of what was supposed to become a leaf. In some ferns at least (e.g., species of *Dryopteris, Hypolepis, Osmunda*), such experimentally induced changes include the developmental switch from a leaf (frond) primordium to a shoot (rhizome) bud and vice versa [176,177].

"Transformation of forms" versus "fixed typology" in the evolution of fern leaves (fronds). Schneider [74] presented a remarkable essay on the evolutionary morphology of ferns (*monilophytes*), while employing a process-oriented approach, which combines the process thinking of the fuzzy Arberian morphology with the process orientation of Darwinian evolution (see Section 3.6). Schneider [74] stressed the heuristic value of continuum plant morphology that employs a holistic view of the plant body in which the different organs, such as leaves, shoots, and roots are linked by shared developmental processes. In comparison to angiosperms, the leaves (fronds) of ferns may share a higher degree of similarity with shoots. Thus, the differentiation between the organs may be understood as less distinct. For example, the leaves of ferns often show a long-lasting apical growth that may be caused by a similar expression of transcription factors in the shoot apical meristems (SAMs) and the leaf apical meristems (LAMs). In addition, several unusual fern structures, for example, stolons of *Nephrolepis*, can be best described as misfits that combine features of different organs [72,74,104,148,174,178,179].

New EvoDevo studies on ferns point to the remarkable similarity of compound leaves (fronds) and shoots [72,104,177]. Raphael Cruz et al. [72] (p. 2) wrote: "Fern leaves are different from most seed plant leaves. For example, unlike seed plants, many fern leaves have a leaf apical meristem (LAM). In ferns, the LAM is responsible for a transient indeterminacy during leaf development, usually producing pinnae during a longer period than the regular compound leaf of a seed plant … Fern leaves resemble the indeterminate shoot by having an apical meristem, producing lateral organs and having a transient or even persistent indeterminacy (as in the genera *Lygodium, Nephrolepis, Salpichlaena, Jamesonia*), as reviewed in Vasco et al. [180]. These features do not fit the classical morphological concept of leaves as they do for seed plants." Based on developmental genetic studies on compound leaves in vascular plants [181], there is evidence that *Class I KNOX* genes are directly associated with indeterminacy and are required to make compound leaves in many cases, representing a partial homology with the shoot. *Class I KNOX* genes are also an important marker of meristematic activity in fern shoots [71]. Cruz et al. [72] found *Class I KNOX* expression in SAMs as well as LAMs (and even in growing pinna tips) of *Mickelia scandens*. Their conclusion: "Fern apical meristems should be interpreted as a complex and highly organized interconnected network of cells with indeterminate fates, specialized zones (apical cells vs. peripheral cells), and the capacity for producing new organs (leaves or pinnae)." According to Cruz et al. [72,104], Harrison and Morris [60], Harrison et al. [148], Plackett et al. [109], and Vasco and Ambrose [177] fern leaves and their segments have to be interpreted evolutionarily and ontogenetically as reduced shoots. Thus, the presence of shoot-like features in developing leaves is strong evidence in favor of Agnes Arber's partial shoot theory (see Section 3.4). Cruz et al. [72] (p. 9) stressed the importance of Arber's ideas with respect to future studies in developmental genetics: "Her theory should be strongly discussed now that new molecular evidence, as our results and other studies discussed here, is available. Our results point to multicellular meristematic structures in the shoot, leaf, and pinna apices, also reinforcing her idea of "identity-in-parallel", in which structures may be put in a relation of the part to the whole, but is also equivalent as a whole. The pinna is part of the shoot, but ultimately is equivalent to a whole shoot, carrying the potential of producing new lateral structures."

4.2. Flowering Plants—Leaf and Shoot Development in Flowering Plants

Compound leaves with determinate apical growth. Foliage leaves in most flowering plants are determinate organs: together with the presence of an axillary bud, this is indeed one of the criteria commonly used to distinguish a "true" leaf from a leaf-like branch (see Section 3.1). Leaves of most flowering plants are initiated sequentially as transversely inserted primordia. They arise at a shoot apical meristem (SAM), which usually shows an indeterminate growth [13,85,176,182]. Similarly, young compound leaves of angiosperm genera such as celery (*Apium*, (Apiaceae), sumac (*Rhus*, Anacardiaceae), neem (*Azadirachta*, Meliaceae), American river-weed (*Marathrum*, Podostemaceae), and orange jasmin (*Murraya*, Rutaceae) are provided with a meristematic leaf tip = leaf apical meristem (LAM). These taxa show an acropetal mode of initiation with leaflet primordia, which are inserted transversely at the LAM. Thus, they resemble leaf initiation at the SAM [49,139,183–187]. — Bharathan and Sinha [188] (p. 1533) summarized while focusing on flowering plants: "The true homologies of compound leaves have been a matter of debate. They have been considered true lateral organs with homologies to simple leaves [189,190], or structures that are intermediate between leaves and shoots [184,191]." Despite the general open-mindedness towards shoot-like features in compound leaves [85,192], there are still various developmental geneticists [193,194] who search for plant hormones and transcriptional regulators modulating compound leaf development—without having the "identity-in-parallel" between leaves and shoots in mind that was formulated by Agnes Arber [28] (p. 142) more than seventy years ago: "To the compound leaf, the leaflet stands in relation of *part* to *whole*, but it is also the equivalent of the compound leaf as a *whole*, though in another generation."

Compound leaves with indeterminate apical growth. Flowering plants have two types of indeterminate leaves: those with a long-lasting apical meristem and those with a long-lasting basal meristem (see below). Indeterminate leaves with a long-lasting leaf apical meristem (LAM), as observable in many ferns (see Section 4.1), exist only in a few groups of flowering plants. Their meristematic activity is maintained for months or even years [56,85,169]. Field studies in woody members of the mahogany family (Meliaceae), especially *Chisocheton* and *Guarea*, have revealed that their compound leaves may show seasonal apical growth for several years. The LAM contributes new leaflets in a manner similar to the SAM [176,191,192,195–197]. Both shoot tips and leaf tips of *Chisocheton* and *Guarea* species show seasonal flushing. Each time the SAM adds new leaves, the LAM also initiates new leaflets. Thus, the meristematic leaf tips and the meristematic shoot tips react to the same endogenous or environmental stimuli. Moreover, the leaf axis (i.e., petiole and rachis) increases in diameter due to secondary thickening with wood production, similar to what is typical for stems (shoot axes) of woody plants. In addition, certain species of *Chisocheton* produce epiphyllous shoots (*Ch. tennis*) or epiphyllous inflorescences (*Ch. pohlianus*) [198]. Thus, compound leaves of *Chisocheton* and *Guarea* have developmental routines resembling whole shoots.

Stipules repeat the developmental pathways of whole leaves. Stipules are basal outgrowths of foliage leaves. Quite often the presence and the typical arrangement of stipules is characteristic for a whole group of plants. Flowering plants such as the rose, pea, and coffee families (Rosaceae, Fabaceae, Rubiaceae) are characterized by stipules [13,192,199–201]. The Plant Ontology Consortium (www.Plantontology.org) provides the following definition for stipule (PO:0020041): "A cardinal organ part that is an appendage at the base of a vascular leaf. Usually one of a pair of appendages. May take the form of a spine (PO:0025174) or may be bristled or brush-like. Stipules may occur on the opposite side of the shoot axis from the leaf they are associated with, but they are still thought to arise from part of the leaf." Leaf blade and stipules usually arise from a common primordial bulge at the shoot apex, but stipules often behave quite independently from the leaf to which they belong. They usually stop growth clearly before the leaf blade itself is mature. They often abscise much earlier than the leaf blade. However, in various flowering plants such as woodruff and allies (*Galium* and related genera in Rubiaceae) leaf and stipule look the same or nearly so, indicating that the stipule is replaced by a leaf. In the flamboyant tree (*Delonix regia*, Fabaceae) the stipules repeat the branching pattern of the much larger compound leaf blade. Various examples of ectopic expression of leaf identity in stipular position are described in [35,192,199]. Arber [28] (pp. 80–82) concluded: "The relation of the stipules to the leaves of which they are members, may be compared with the relation of cotyledons to the primary shoot, and of prophylls or bracteoles to a lateral shoot."

Developmental genetics may help in defining "leaf identity" versus "stipule identity". A homeotic replacement of the stipule by a leaf was described in pea (*Pisum sativum*) mutants such as *cochleata* [202]. There are tendrils forming the blades as well as tendrils arising from stipular positions, a situation not known from any wild-type member of Fabaceae [129,152,203]. In the *afila* (af) mutant, all primary pinnae are replaced by a bunch of tendrils, whereas the stipules are not altered. A gene known to interact with the *af* gene is *sinuate leaf* (*sil*), which results in undulating margins of both leaflets and stipules. When combined with *af, sil* plants have adventitious tendrils arising from clefts in the distal portion of the stipule [35,203]. Thus, pertinent characters of the leaf blade can be ectopically expressed in stipular sites. Yaxley et al. [202] wrote that these mutants "change stipules into a more 'compound leaf-like' identity".

Stipules in flowering plants are, by definition, restricted to the leaf base. However, a few mutants in Arabidopsis and pea are known to have supernumerary stipules, which are ectopically expressed as part of the leaf blade or rachis [204]. The so-called "stipels" at the base of the lateral leaflets in compound leaves of the garden bean (*Phaseolus vulgaris*) may be understood as ectopically expressed supernumerary stipules, repeating the developmental leaf program at the base of its subunits [13,28].

Indeterminate meristematic activity at the leaf base in one-leaf plants (*Monophyllaea*, *Streptocarpus*). Besides leaves that grow indeterminately at their tip (as typical for ferns and few flowering plants) there are one-leaf plants having only one large foliage leaf at maturity. Because of this bauplan oddity, they were called "morphological misfits" [75] (see Section 3.2). The only foliage leaf per plant equals one of the two cotyledons that grows indeterminately due to the long-lasting meristematic activity at the base of the lamina [56,85,205,206]. One-leaf plants mainly belong to the genera *Monophyllaea* and *Streptocarpus* in the Gesneriaceae, which is a family of asterids in eudicots [207]. Unlike typical vascular plants, one-leaf plants do not have a typical shoot apical meristem (SAM).

The developmental biology of the one-leaf plants (*Monophyllaea, Streptocarpus* spp.) was recently studied by Kinoshita and Tsukaya [69,208]. The indeterminate growth of the only foliage leaf in these plants is supported by a groove meristem. Gene expression and physiological analyses proved that the groove meristem is equivalent to the SAM of typical flowering plants. Using classical terms for structural units in the one-leaf plants may be problematic. In *Streptocarpus,* there are several species with a usual bauplan, with elongate stems carrying several paired leaves. Somewhat comparable with the phytomeric shoot model (see Section 3.3), Jong and Burtt [205] proposed the phyllomorph concept fifty years ago in order to describe the unusual structure of the one-leaf plants (e.g., *Streptocarpus wendlandii*) lacking an obvious stem. The "phyllomorph" is defined as a novel but fuzzy developmental module consisting of an indeterminately growing leaf lamina and a "stem- and petiole-like structure" (called petiolode) at its base. Thus, the phyllomorph of a one-leaf plant corresponds to a shoot (including stem) in typical plants [69,85,208]. For a recent review on *Streptocarpus* phyllomorphs including regulation of meristem activity by plant hormones, see Nishii et al. [209].

"Stem-leaf mixed organs" in river-weeds (Podostemaceae). River-weeds are peculiar flowering plants that are adapted to river-rapids and waterfalls in the tropics. They are related to St John's wort (*Hypericum*) within the eudicots. Bell [75] called the river-weeds "morphological misfits", because they follow structural rules that are different from those of more typical flowering plants (see Section 3.2). The boundaries between organs known as roots, stems, and leaves—distinguishable in related flowering plants – are blurred (fuzzy). However, the members of this family have stable floral bauplans [22]. When we cling to structural categories such as leaf, stem, and root for the description of the plant bodies in river-weeds, we get into trouble with either/or homology ("sameness") of the various plant parts. Then we are forced to accept the existence of structural intermediates such as "stem-leaf mixed organs" in Podostemaceae, as genetically analyzed by Katayama et al. [82,83]. This appellation is meant to indicate that these structures have some features of leaves, and some of stems, probably due to their unusual gene expression pattern and the lack of obvious SAMs [84]. Summarizing the developmental genetic results in Podostemaceae and other flowering plants, Eckardt and Baum [210] concluded: "It is now generally accepted that compound leaves express both leaf and shoot properties."

4.3. Flowering Plants—Bladderworts and Allies (Lentibulariaceae)

"It is probably best, as a purely provisional hypothesis, to accept the view that the vegetative body of Utricularias partakes of both stem nature and leaf nature. How such a condition can have arisen, historically, from an ancestor possessing well-defined stem and leaf organs, remains one of the unsolved mysteries of phylogeny." Agnes Arber (1920, p. 107) [211]

The case studies above (Sections 4.1 and 4.2) have shown that leaves of vascular plants can have indeterminate growth. They can even have terminal buds. This case study covers the bladderworts (*Utricularia* and allies), plants with highly unusual architectural rules deviating considerably from the body plan of typical flowering plants. Minelli [56] (p. 254) summarized the puzzling situation found in bladderworts properly: "Mechanisms specifying the identity of individual organs are unusually plastic: of the various primordia in a rosette, some will become leaves, others stolons, but their fate can be changed even at a very late stage of development. In some species, a stolon can eventually stop growing, completing its activity with the production of a terminal leaf. On the other hand, the tip of a nearly mature leaf can continue growing and turn into a new stolon." It is difficult to describe and interpret the developmental architecture of the bladderworts in terms of classical plant morphology. Continuum approach and process thinking provide complementary perspectives if we really want to understand the morphological oddities in this group (see Sections 3.5–3.7). In their flowers, however, the bladderworts behave as typical flowering plants, showing that they belong to the asterid eudicots [22,75,212].

The Lentibulariaceae are a family of flowering plants showing carnivory. They attracted and still attracts plant scientists, mainly because of the unique features in biology, morphology, and developmental genetics. The three genera in this family can be distinguished by their specialized traps for carnivory:

flypaper traps (passive) in *Pinguicula*, eel traps = lobsterpot traps (mainly passive) in *Genlisea*, and bladder traps (that conduct suction in < 1ms) in *Utricularia*. The genus *Utricularia* (bladderworts) is with >230 species by far the largest of the three genera. These plants live fully immersed in water (*U. australis, U. gibba, U. vulgaris*), on wet soil (e.g., *U. dichotoma, U. livida, U. reniformis, U. sandersonii, U. longifolia*), or even epiphytic (e.g., *U. alpina*, again *U. reniformis*) [212–214]. Irrespective of their ecological preferences, all *Utricularia* species are provided with tiny bladders used as active sucking traps for small animal prey in water.

The Lentibulariaceae have attained recent attention because of their dynamic nuclear genome size. Their genomes span an order of magnitude and include species with the smallest genomes in angiosperms, making them a powerful system to study the mechanisms of genome expansion and contraction. Moreover, developmental geneticists started to reveal the genetic network behind the morphogenesis of the peculiar traps in *Utricularia* [215,216]. Thus, there is a strong need for plant morphologists and biophilosophers to complement the research progress in developmental genetics of *Utricularia* and allies. There is a rich literature on comparative development and morphology in Lentibulariaceae provided by botanists dating back more than a hundred years (as summarized in [13,22]).

Leaves, shoots, and stolons as descriptive terms for bladderworts. Peter Taylor [212], in his marvelous book on bladderworts taxonomy (genus *Utricularia*), used structural terms such as leaves, stolons, and air shoots for the description of the vegetative bodies in bladderworts. He was well aware that a clear-cut and consistent terminology in plant morphology is needed to become understandable for a reading audience but may not be adequate to the reality. Taylor [212] (p. 6) wrote about plant architecture in bladderworts: "Quite apart from the traps themselves the organization of the vegetative organs of *Utricularia* is peculiar and quite different from that of other flowering plants... For taxonomic and descriptive purposes, whatever their true or theoretical nature, it is desirable to have a consistent terminology for the various organs. Such a terminology has evolved, through the work of Goebel, Lloyd, and others. Yet, because of their great diversity and often plastic nature, it is not always easy to apply the terms in a consistent manner, but for practical purposes, they are found to serve in the majority of cases." Taylor [212] (p. 8) continued about the strange *Utricularia* bauplan with a kind of British humor, while pointing to Rutishauser and Sattler [32]: "It seems reasonable to me that plants which have, in defiance of the general rule, no radicle or roots, may be allowed to have leaves which also disobey the rules. By calling them leaves a few difficulties do arise, but these are relatively unimportant in the context of the overall diversity and nonconforming nature of the vegetative morphology as a whole".

Species of *Utricularia* show various examples of **developmental mosaics** between structural categories typically referred to as leaf and shoot (see above). In certain aquatic members such as *U. purpurea*, the developmental pathway of the whole shoot is repeated within each compound leaf to an astonishing degree [199]. This, together with other observations, led Goebel [217] to the somewhat exaggerated conclusion that a primordium in *Utricularia* can grow into any organ such as a trap, leaf, green shoot (i.e., leafy stem), anchorage shoot, or inflorescence. This generalization is only partly accurate. The developmental and positional constraints in *Utricularia* deviate considerably from the rules of classical plant morphology. *Utricularia*'s plant body, thus, is better understood within the conceptual framework of continuum and process plant morphology [13,18,22]. Based on genomic data, the developmental geneticists Silva et al. [214] (p. 15) presented the following evolutionary hypothesis with respect to the two model species: "*Utricularia gibba* seems to have a more severe degree of fuzzy Arberian morphology, such as no clear delimitation of distinct vegetative organs. In contrast, *U. reniformis* presents a more traditional vegetative organ delimitation (as stems and leaves), similar to other angiosperms."

Members of the morphological school of Rolf Sattler [18,34,138,156] as well as Reut and Plachno [218] represented the vegetative bodies of aquatic and terrestrial bladderworts (e.g., *U. foliosa, U. dichotoma*) and other morphological misfits as combinations of developmental processes using multivariate statistical analyses. Plant organs such as watershoots, leaves, or bracts in *U. foliosa*, also various kinds of stolons in *U. dichotoma* are identified in the morphospace as specific process combinations (see Sections 3.6 and 3.7). No doubt, the use of process combinations to describe plant structures makes communication among scientists difficult. Nevertheless, one of the great strengths of this approach is that categorical terms such as leaf and shoot serve only as placeholders for combinations of developmental processes that locate the organs in the morphospace. Gene expression patterns of model bladderworts (such as *U. vulgaris*, *U. gibba*) and related *Genlisea* species may finally be annotated to the morphospace by associating them with the combination of processes that are found in the parts where the genes are expressed [34,216,219,220].

How to avoid inconsistencies regarding the seeming lack of roots in *Utricularia?* Difficulties in distinguishing "root identity" and "shoot identity" (leafy stolons) are known from Lentibulariaceae, such as bladderworts (*Utricularia*) and butterworts (*Pinguicula*):

Scenario A = 'Loss-of-Root' hypothesis: Usually it is said that *Pinguicula* has roots, and the sister genera *Genlisea* and *Utricularia* lack them. Most botanists and developmental biologists insist that only the more basal genus *Pinguicula* still has true roots, whereas the members of the more derived *Genlisea–Utricularia* lineage have lost them completely [214,216,219–223]. Peter Taylor [212] (p. 6) came up with the following conclusion for *Utricularia*: "Roots are always absent, but organs that resemble and function as roots (here termed rhizoids) are usually present."

A main argument in favor of this scenario is the loss of various root-specific genes in *Utricularia gibba*, as summarized by Renner et al. [223]. According to them, various genes responsible for the growth of roots and root hairs are lacking (or apparently so) in *U. gibba*. Renner et al. [223] summarized (p. 147): "Taken together, loss of genes important for root and shoot morphogenesis may be correlated with the absence of an obvious root in *U. gibba* and could also help to define its unique bauplan suited to an aquatic habit."

Scenario B = 'Root–Stolon Transformation' hypothesis: *Utricularia* stolons may have arisen from what are called "roots" in *Pinguicula* just by adding exogenous leaves to the root surface, as already proposed by Brugger and Rutishauser [224]. This would explain the high degree of similarity of the stolons (stems) of various bladderworts with *Pinguicula* roots. Stolons of various terrestrial *Utricularia* species show an "awkward" phyllotaxis pattern (known as orthomonostichy) with all leaves arranged along a single stem sector [13,22,34,218]. According to the continuum plant morphology, the roots were not completely lost in the *Genlisea–Utricularia* lineage. The ancestral roots (as still present in *Pinguicula*) evolved exogenous green appendages that can be called "leaves" (an idea anticipated a hundred years ago by Arber [211]). Thus, the developmental pathways for roots and shoots were blended (amalgamated) to some degree, perhaps due to co-option of genes usually acting in stems and leaves but not in roots. Arguments in favor of this "root–stolon transformation" hypothesis are as follows: (i) Several *Pinguicula* have roots without caps (e.g., *P. moranensis*) [13,34]. (ii) Various Utricularias (e.g., *U. longifolia*, *U. livida*, *U. sandersonii*) have straight stolon tips which look (including anatomy) similar to cap-less root tips of *Pinguicula*. When *U. longifolia* is cultivated in a hanging pot, the root-like stolons breaking through the bottom show positive geotropism. (iii) Conversion of root meristems to shoot meristems are also known from other angiosperms such as *Nasturtium* (Brassicaceae) and *Neottia* (Orchidaceae), pointing to a morphological correspondence between root and shoot (as discussed by Guédès [225]).

Many developmental genes involved in the morphology of the *Genlisea–Utricularia* lineage were uncovered within the past few years. Genome analyses of *G. aurea*, *U. gibba*, and *U. vulgaris* (all of them apparently rootless) showed the presence of a considerable number of root-specific genes in their vegetative bodies [219,222,226,227]. For instance, genes involved in the root hair formation are present in *Utricularia*. Thus, the absence of a typical root can be a result of the expression (or lack thereof) of specific transcription factors and, therefore, not a result of lacking root developmental genes (Miranda, pers. comm.).

The 'loss-of-root" hypothesis (A) and the "root–stolon transformation" hypothesis (B) in the *Genlisea–Utricularia* lineage may merge into one when developmental processes and gene actions are emphasized instead of arbitrary structural categories. In the context of process morphology (getting rid of structural categories), both seemingly contradictory views are complementary perspectives of the same reality. The apparent loss of roots in *Utricularia* and *Genlisea* as compared to *Pinguicula* is a wonderful paradoxon where two contradictory perspectives express the same developmental reality. Both root and shoot meristems of vascular plants are regulated by common genetic mechanisms [223,228,229]. Rcut and Pladuio [218] (p. 17) hypothesized the following about stolon evolution in *U. dichotoma* as basal member of the genus *Utricularia* (subgenus *Polypompholyx*): "In terms of a continuum or morphocline of process combinations, and in the phylogenetic context, developmental processes for 'shoot' (e.g., branching) and developmental processes for 'leaf' (e.g., a determinate growth period) may have been added to a 'reduced *Pinguicula* root'." However, the *Utricularia* stolons should not be labelled as "obvious roots" or even "true roots". We should only keep in mind scenario B as complementary perspective: What is usually called "stolons" in *Utricularia* may have evolved from roots (cf. *Pinguicula*) that started to produce exogenous leaves.

Somewhat comparable to Arber's partial-shoot theory of the leaf (Sections 3.4 and 4.1), roots and shoots (stems, stolons) of vascular plants in general may be partially homologous, as summarized by Minelli [56] (p. 248): "From a phylogenetic perspective, the Continuum Model is consistent with the hypothesis that both leaves (e.g., [228,230]) and roots (e.g., [231,232]) are derived from shoot-like organs, at least in ferns and seed plants." Arber [86,103] already proposed the "partial-shoot theory of the root", coming close to the leaf-skin shoot model [97] (see Section 3.3), with the stem core being equivalent to the "root" and the stem cortex being formed by the elongated leaf bases, which cover the stem core like a skin. Arber [103] (p. 101) was aware of possible flaws of such a view when she wrote: "The supposition that the root may represent the 'core' of the shoot is not intended to imply anything so naively simple as that the root is anatomically a continuation of the central region of the shoot alone; this, indeed, would be obviously untrue, at least for the main root. It is held, rather, that the root is endowed with tendencies and capacities which correspond to those of the inner region of the shoot rather than those of its surface."

4.4. Flowering Plants—Flowers, Inflorescences, and Intermediates

The origin of flowers was a key innovation in the history of seed-plants between 150 and 190 million years ago. Typical flowers in flowering plants (angiosperms), especially in the large group of eudicots, consist of four whorls with sepals (usually green), petals (often showy and colorful), stamens (for pollen production), and carpels (often fused into one unit, the ovary). The latter contain the ovules which—once fertilized—become the seeds as result of sexual reproduction. The present case study does not allow us to consider all the wonderful morphological types of flowers and pollination syndromes in the c. 300,000 species of angiosperms, as presented in books and book chapters by, e.g., Endress, Johnson and Schiestl, Minelli, Ronse De Craene, Sattler, and Soltis et al. [56,233–237].

The developmental geneticists Elliott Meyerowitz and Enrico Coen (see e.g., [238]) proposed the ABC model of floral organ specification, mainly based on the two model plants Arabidopsis (thale cress, Brassicaceae) and Antirrhinum (snapdragon, Plantaginaceae). More recently, the ABC model turned into the more elaborated ABCE model. It explains the bauplan of typical flowers with four consecutive whorls (sepals, petals, stamens, carpels) by serial expression of organ identity genes, especially MADS box genes [124,235,239–243]. After having successfully explained the organ position in model organisms such as Antirrhinum and Arabidopsis, the developmental geneticists went on to explain flowers which form continua between sepals, petals, and stamens (in e.g., *Nymphaea*, waterlily). Thus, Soltis et al. [235,242] offered a **"fading borders" model** as a testable hypothesis for these floral organs of intergrading morphologies in basal lineages of flowering plants such as waterlilies, where the B-function genes are more broadly expressed. In the "fading borders" model, there is a gradual transition of gene function across floral organs. This is unlike the classical ABCE model, where there is a sharp boundary between gene expression corresponding to floral organs [241,244–246].

Developmental mosaics between sepals, petals, and stamens are accepted by proponents of both classical and continuum plant morphology (see Section 3.2), because these kinds of floral appendages are understood—since Goethe—as modified leaves (phyllomes) [61,92,105]. Arber [28] (p. 55) summarized the history of this idea with emphasis on petals and stamens: "We may indeed agree with Goethe (1790) [92] and de Candolle (1827) [247] that petals and stamens show so much affinity that it is evidently reasonable to group them together. The petals will then be regarded as transition members between the vegetative and the actively reproductive parts of the floral shoot."

Intermediates between flowers and inflorescences. Still questionable in botany is the morphological significance of the angiosperm flower. Claßen-Bockhoff and Frankenhäuser [132] (p. 8) confessed: "While intermediate structures or even 'hybrids' between flowers and inflorescences are widely accepted (e.g., Prenner and Rudall [248]; Kirchoff et al. [34]), their morphogenetic and phylogenetic significance is not yet fully understood." Although many reproductive structures are easily identified as flowers, there are some examples lying in-between flowers and inflorescences, e.g., the "terminal flower-like structures", as observable in basal angiosperms and early monocots with elongate inflorescences [249].

Consider for example the staminate unit ("male flower") of the castor oil plant (*Ricinus communis*) that is known for its branched "staminal trees". Each unit may be seen as a single male flower with multiple branched stamen-fascicles [35,248,250,251]. Claßen-Bockhoff and Frankenhäuser [132] presented a somewhat contradicting view: Each "male flower" (branching staminate unit) of *Ricinus communis* is interpreted as a multi-flowered unit composed of highly reduced uni-staminate flowers, similar to what is known from *Euphorbia* within the same family as *Ricinus*.

Gene expression patterns are used to distinguish between flowers and inflorescences in Euphorbiaceae to which *Ricinus* belongs. Prenner et al. [251] found the expression of the LEAFY (LFY) protein not only in individual flower primordia of *Euphorbia*, but also in the primordium that leads to a group of flowers (called "cyathium"), indicating that inflorescence meristems may be genetically similarly regulated like floral meristems. Additional studies in various flowering plant families illustrate that one should be careful using gene expression patterns for floral organ homology [115,246,252].

Claßen-Bockhoff and Frankenhäuser [132] proposed floral unit meristems (FUMs) as intermediate stages between inflorescence and flower meristems: "Ongoing fractionation defining the transition from flowers to floral units may be based on a delay in the expression of floral organ identity genes." Claßen-Bockhoff and Frankenhäuser [132] remained vague with respect to male reproductive units (inflorescences, flowers?) in *Ricinus*. They concluded: "If the staminate unit is nevertheless interpreted as a flower (which is possible), one has to accept a completely new stamen structure and to disregard the meristem similarities between *R. communis* and floral units. Maybe, it is too early for a final morphological interpretation of the staminate units in *Ricinus,* as there is too little known about FUMs (=floral unit meristems), their diversity, and genetic regulation." FUMs allow a new view on intermediate stages/structures between flowers and inflorescences, because all types of reproductive meristems originate from shoot apical meristems. Somewhat similar fractionation steps as found in *Ricinus* reproductive development are observable inside the branching flowers of the tropical waterlily *Nymphaea prolifera*, leading to a multitude of sterile "daughter flowers" and "granddaughter flowers" that serve as vegetative propagules [34,253].

Shoot-bearing leaves in *Hooded* barley. Epiphyllous flowers are known from several angiosperms [95]. The developmental genetics behind this type of ectopic flower position is not yet fully understood. There is one nice example that was described and illustrated by Agnes Arber in her book on grasses [254]: the shoot-bearing leaf in Nepal barley (*Hordeum vulgare* var. *trifurcatum*). It shows epiphyllous spikelets arising from the awned bract (called "lemma" in grasses). By examining their relative positions, Arber [254] (p. 312) concluded that the lemma (bract) "which is a *leaf* member, behaves to the accessory spikelet in all respects as if it were that spikelet's parent *axis* [her italics]." The very same situation is found in the *Hooded* mutant of *Hordeum vulgare* which was described later [255,256]: In *Hooded* barley plants, one or more extra flowers (spikelets) develops at the site of transition between bract ("lemma") and awn. Molecular studies [257–259] have elucidated that the *Hooded* phenotype of barley is caused by a duplication in a homeobox gene intron. Williams-Carrier et al. [259] suggested that the inverse polarity of the ectopic spikelets seen in the *Hooded* mutant of barley results from the homeotic transformation of the lemma awn into a reiterative inflorescence axis. This is an example of conversion of organ identity: a leaf part (the awn) is converted into a shoot axis. Neither Arber's book on grasses [254] nor the *Hooded* barley are mentioned in a recent review [260] on the developmental genetics of grass inflorescences.

4.5. Phyllotaxis: The Algorithmic Beauty of Plants

Shoot apical meristems (SAMs) in vascular plants give rise to organs at their flanks in a periodic and predictable pattern. Phyllotaxis is the spatial arrangement of lateral organs along a stem, within a rosette or shoot bud, or in a flower. The repeated lateral organs involved are usually leaves or flower parts. The study of phyllotaxis covers pattern formation in and around SAMs of vascular plants. Some phyllotactic patterns are easy to observe and characterize whole groups of plants. For example, leaf whorls with several units emerging from the same node are typical for horsetails (*Equisetum*), a group of flowerless and seedless vascular plants related to ferns [56,199]. Another group of regular phyllotactic patterns are the spiral ones, when the young leaves (or floral parts) arise one by one around the periphery of the SAM. For the past two hundred years, botanists and mathematicians tried to answer the question why most spiral patterns in seed plants approach the Fibonacci angle (c.137.5°), dividing the full circle (360°) according to the Golden section [168,261–267]. Barabé and Lacroix [268] summarized historical aspects and new research trends in a concise book entitled "Phyllotactic Patterns—a Multidisciplinary Approach". Phyllotaxis research applies mathematical models on pattern formation without bothering too much about the morphological significance of the plant organs involved [269]. In most cases, the lateral structures (appendages) arising at the periphery of SAMs are leaves, or axillary flowers, subtended by bracts in, e.g., the flower-heads of Asteraceae. There are only few cases known where leaves are replaced by lateral shoots or flowers at SAMs, e.g., in waterlilies (Nymphaeaceae) [253,270,271]. Natural examples of shoot or flower development from primordia occupying leaf sites are also known in clubmosses (*Huperzia*) [272,273], and *Utricularia* (see Section 4.3).

Process thinking and the continuum approach seem to be quite useful in phyllotaxis research. For example, structural units of seemingly unequal morphological significance (e.g., leaves, leaflets, and stipules) can be involved in the formation of leaf whorls of various flowering plants [35,199,200,266]. Leaf rosettes with attractive Fibonacci spiral patterns can even result from repeated sympodial branching (i.e., determinate shoot modules) combined with flowering, as found in *Pinguicula* [274]. Moreover, nature is not always as orderly as we think. There are many examples of vegetative and floral meristems where spirals and whorls are replaced by more irregular or even chaotic patterns, including clustered leaves on one side of the SAM. For example, irregular (chaotic) initiation of c. 200 stamens is observable in flowers of ylang-ylang (*Cananga odorata*, Annonaceae) [200,266,275].

Barabé and Lacroix [268] describe the interplay of biophysical parameters (e.g., tissue tensions in and around the SAM), size ratios between SAM and incipient leaves, and endogenous factors such as phytohormones (especially auxin) and gene expression that seem to be responsible for the emergence of regular patterns [261]. Molecular developmental geneticists already detected several genes that are involved in the emergence of disordered or perturbed patterns. Phyllotactic mutants, however, do not directly influence the size of the divergence angles (e.g., stronger deviations from the Fibonacci angle). They mainly change the three-dimensional shape and size of the SAM as the site of self-organizing processes [261,268,276,277]. During plant growth, including size increase of the SAM, various transitions between whorled, intermediate, and spiral patterns may occur, also depending on the taxon involved [263,265,278]. Clubmosses (*Lycopodium* and allies), for example, switch more easily between patterns that are rare to very rare in seed plants [273,279,280]. Dynamic models and computer simulations based on process thinking, similar to those already presented for transitions in phyllotaxis, will help in the near future to better understand the complete architecture of vascular plants [155,168,268].

Leaf initiation without an obvious SAM: Finally, we should keep in mind that a shoot apical meristem (SAM) is not a "sine qua non" for leaf inception in flowering plants: (i) There are Australian *Acacia* spp. (e.g., *A. verticillata, A. baueri*) with "leaf whorls" that increase the number of whorl members by initiation of additional leaves (phyllodes) in meristematic stem zones below the proper SAM, as already described by the famous Wilhelm Hofmeister in 1868 [199,266,281]. (ii) The river-weeds Podostemaceae (adapted to river rapids and waterfalls in the tropics), especially its subfamily Podostemoideae, reveals leaf initiation from the adaxial side of an already formed young leaf. The shoot lacks a recognizable SAM with permanent stem cells. The SAM is cryptic or even lacking. A new leaf primordium in a vegetative shoot tip develops from the base of the opposing second youngest leaf primordium. The initiation of a new leaf primordium appears to be associated with degeneration of neighboring cells, as shown by Japanese botanists [82–84,282]. This kind of leaf formation without SAM is repeated, resulting in a chain of leaves or phytomeres (see Section 3.3).

5. Conclusions: Various Ways to Express Plant Growth and Architecture as Process Combinations and Developmental Continua

A sentence credited to the Greek philosopher Heraclitus of Ephesus (c. 535 BC–475 BC), "everything flows—nothing stands still", was used many times as a metaphor in lyrics and common speech. Heraclitus probably was one of the first process philosophers [38,39]. "Everything flows" is also the title of a book [33,40,283] as well as a paper [284] on processual philosophy of biology. The present essay focused on the philosophical foundations of process thinking and continuum views in biological sciences, as summarized by Jaeger and Monk [284]: "The process perspective holds that reality is fundamentally dynamic: the basic constituents of the universe are interconnected sequences of occurrences or events. Accordingly, reality must be described and explained using explicitly dynamic, that is processual, concepts rather than notions representing static 'things' or entities."

As a botanist with a passion for plant morphology (the main issue of the present essay), I have to admit that many interesting aspects of process philosophy could not be discussed here, including the heuristic value of continuum views in biological disciplines beyond plant morphology. For example, other biological disciplines such as systems biology, neurobiology, and cognitive sciences seem to be far ahead of plant biology/morphology with respect to process thinking. Jaeger and Monk [284] presented an example: "It is impossible to think about the nerve impulses that provide the substrate for thought, and ultimately the higher-level phenomena of mind and consciousness, in any other than dynamic terms." Also not covered by my essay are morphogenetic fields as dynamical modules and promising steps towards dynamical systems biology [16,17,285].

With respect to plant sciences, Sattler [6] (p. 61) admitted: "Although it seems easy to say that everything flows (changes), to arrive at a completely dynamic view of plant morphology is not an easy task." The present essay emphasizes the need for complementary perspectives, continuum, and process thinking in EvoDevo of both plants and animals. Historical roots of this open-minded way to look at growing organisms, especially plants, go back to Johann Wolfgang von Goethe, Agnes Arber, and Rolf Sattler. Multicellular plants such as seed plants, ferns, and bryophytes usually show algorithmic growth with continued branching and repetition of modules. The acceptance of growing plants as dynamic continua allows developmental geneticists and evolutionary biologists to move towards a more holistic understanding of plants in time and space [1,273].

There are several ways to understand the complexity, continuity, and dynamics of plant growth and architecture:

(i) Recognize examples of unusual morphologies in plants by describing them as **morphological misfits** (see Section 3.2, Sections 4.1–4.4). Flowering plants with bauplans that deviate strongly from the approach of classical plant morphology were labelled as "morphological misfits" by Bell [75]. Being a misfit is not the problem of the plant, but the problem of our inadequate thinking and concepts. Morphological misfits do not fit classical plant morphology, which, however, is still useful as a rule of thumb in many usual (or normal) groups of flowering plants.

(ii) Accept **scientific perspectivism** and complementary ways of describing the same plant in space and time: Stress that one kind of interpretation is often not enough to explain the architectural complexity. For example, the modular growth of a leafy shoot may be described by the phytomere model as well as by the traditional shoot–leaf model (see Section 3.3) [32].

(iii) Consider the **principle of iteration** in plant development. A subunit such as a foliage leaf repeats to some degree the development of the whole shoot. This view coincides with the identity-in-parallel concept and the "partial-shoot theory" of Agnes Arber [28,49,72]. It has its counterpart in the anchor concept in zoology, where the paramorphism concept was proposed by Minelli [106,107] for multicellular animals such as tetrapods: Animal appendages can be regarded as a partial repetition of the main body axis (see Section 3.4).

(iv) Use **process morphology** and mathematical tools to define intermediacy between typical plant organs: Process thinking and the continuum approach in plant morphology allow us to perceive and interpret growing plants as developmental continua, as process combinations rather than as assemblages of structural units ("organs"), such as roots, stems, leaves, and flowers. Therefore, we may use strict sets of developmental processes for defining leaves vs. stems vs. roots. Then let us code these characteristics and use statistical tools like principal component analysis (PCA) for the distinction of typical plant organs (as found in many vascular plants) and developmental mosaics between structural categories. Homology includes partial homology and quantitative homology, as proposed by Sattler [6,113,119,133]. In vascular plants, this leads to a continuum between structural categories (plant organs) such as roots, stems, leaves, and even multicellular hairs/trichomes. How intermediates between typical plant organs are best described has been discussed by Kirchoff et al. [34] and Lacroix et al. [49]. For example, the structures observed in the bladderworts (*Utricularia*) may be called leaf and stem for convenience, as done by Taylor [212], although other authors [22,218] interpreted them as developmental mosaics including root components (see Section 4.3).

(v) Accept **developmental genetics** as the 4th homology criterion for defining the morphological significance of unusual plant structures. Traditionally, three homology criteria were used: position, special quality, and the existence of intermediates [62,108,113,117,118]. Unlike the more holistic Sattler school, reductionistic biologists such as Scheres et al. [286] (p. 963) emphasized the primacy of molecular genetics over traditional morphology/anatomy: "Regardless of how much faith one has in anatomical definitions, they should not be taken as more than a means of communication prior to subsequent genetic analysis." Similarly, developmental geneticists may insist on the primacy of organ-identifying genes over the three traditional homology criteria [125]. For example, by stating that the bladderworts (*Utricularia*) lack important genes for roots, there is a genetic basis for the lack of (typical) roots in the bladderworts [223] (see Section 4.3).

(vi) Accept **developmental processes** such as homeosis, ectopic expression, blurring, and upgrading of organ identities in plant structures that transcend typical bauplans. Although the concept of homeosis is much older than developmental and molecular genetics, it gained much additional explanatory power with the discovery of homeotic genes (organ-identity genes), such as the MADS box genes explaining the bauplan of typical flowers in angiosperms [119,124] (see Section 4.4). When classical bauplan rules of vascular plants (consisting of roots, stems, and leaves) are violated, then it becomes difficult to clearly define and discriminate between the three types of organs. Rutishauser et al. [35] described several examples of plants "having identity crises".

Identity crises result from our inadequate vocabularies while describing and interpreting plant architecture in space and time.

(vii) Design **virtual plants** using iterated developmental processes. The development of new mathematical concepts and computational techniques for the description of growing plant structures can be based on developmental rules such as branching, repetition of growth units (e.g., phytomeres), and environmental parameters, as already done by Prusinkiewicz and colleagues [155,168,267,276,287].

(viii) Maybe we need to do more of what Agnes Arber and other morphologists since Goethe have done: Rely more on the visual abilities that draw people to biology. Get a **feeling for the organism** [8–12,61,134,288,289].

(ix) Last but not least: **Celebrate** your achievements towards process and continuum thinking with a sip of Agnes-Arber Gin: "A fantastic gin celebrating the renowned botanical historian Agnes Arber". **Cheers!** https://agnesarbergin.signature-brands.co.uk/online/—In addition, you may listen to the song on EvoDevo (Despacito Biology Parody), a music video performed by the Canadian science communicator and youtuber Tim Blais (A Capella Science): https://www.youtube.com/watch?v=ydqReeTV_vketlist=PL20YbtNRgutzZftYyTI_p2G3ttd4R9dVs.

Funding: This research received no external funding.

Acknowledgments: I am grateful to Alessandro Minelli for inviting me to contribute to the special issue on "Renegotiating Disciplinary Fields in the Life Sciences" and for commenting on a draft version of this contribution. I would also like to thank Denis Barabé, Volker Bittrich, Christian Lacroix, Rolf Sattler, and the two anonymous peer reviewers for their criticisms, comments, and suggestions.

Conflicts of Interest: The author declares no conflict of interest.

References

1. Sattler, R.; Rutishauser, R. The fundamental relevance of morphology and morphogenesis to plant research. *Ann. Bot.* **1997**, *80*, 571–582.

2. Pavlinov, I.Y. Multiplicity of Research Programs in the Biological Systematics: A Case for Scientific Pluralism. *Philosophies* **2020**, *5*, 7. [CrossRef]

3. Amato, S.I. EvoDevo: An Ongoing Revolution? *Philosophies* **2020**, *5*, 35. [CrossRef]

4. Minelli, A. Disciplinary Fields in the Life Sciences: Evolving Divides and Anchor Concepts. *Philosophies* **2020**, *5*, 34. [CrossRef]

5. Sattler, R. *Biophilosophy. Analytic and Holistic Perspectives*; Springer: Berlin/Heidelberg, Germany; New York, NY, USA; Tokyo, Japan, 1986.

6. Sattler, R. Structural and dynamic approaches to the development and evolution of plant form. In *Perspectives on Evolutionary and Developmental Biology. Essays for Alessandro Minelli*; Fusco, G., Ed.; Padova University Press: Padova, Italy, 2019; pp. 57–70.

7. Sattler, R. Beyond Wilber. Available online: beyondwilber.ca (accessed on 28 November 2020).

8. Flannery, M.C. Agnes Arber: Form in the mind and the eye. *Int. Stud. Philos. Sci.* **2003**, *17*, 281–300.

9. Flannery, M.C. Spotlight. Agnes Arber in the 21st Century. *Systematist* **2005**, *24*, 13–17.

10. Elkin, R.S. Live matter. Towards a theory of plant life. *J. Landsc. Archit.* **2017**, *12*, 60–73.

11. Kirchoff, B.K. Preface: From Agnes Arber to new explanatory models for vascular plant development. *Ann. Bot.* **2001**, *88*, 1103–1104.

12. Kirchoff, B.K. Character description in phylogenetic analysis: Insights from Agnes Arber's concept of the plant. *Ann. Bot.* **2001**, *88*, 1203–1214.

13. Rutishauser, R.; Isler, B. Developmental genetics and morphological evolution of flowering plants, especially bladderworts (*Utricularia*): Fuzzy Arberian Morphology complements Classical Morphology. *Ann. Bot.* **2001**, *88*, 1173–1202.

14. Sattler, R. Some comments on the morphological, scientific, philosophical and spiritual significance of Agnes Arber's life and work. *Ann. Bot.* **2001**, *88*, 1215–1217. [CrossRef]

15. Claßen-Bockhoff, R. Plant morphology: The historical concepts of Wilhelm Troll, Walter Zimmermann, and Agnes Arber. *Ann. Bot.* **2001**, *88*, 1153–1172. [CrossRef]
16. Lander, A.D. The edges of understanding. *BMC Biol.* **2010**, *8*, 40. [CrossRef] [PubMed]
17. Jaeger, J. The importance of being dynamic: Systems biology beyond the hairball. In *Philosophy of Systems Biology. Perspectives from Scientists and Philosophers*; Green, S., Ed.; Springer: Cham, Switzerland, 2017; pp. 135–146.
18. Sattler, R.; Rutishauser, R. Structural and dynamic descriptions of the development of *Utricularia foliosa* and *U. australis. Can. J. Bot.* **1990**, *68*, 1989–2003. [CrossRef]
19. Sattler, R. Classical morphology and continuum morphology: Opposition and continuum. *Ann. Bot.* **1996**, *78*, 577–581. [CrossRef]
20. Korzybski, A. *Science and Sanity: An Introduction to Non-Aristotelian Systems and General Semantics*, 5th ed.; Institute of General Semantics: Fort Worth, TX, USA, 1994.
21. Giere, R.N. *Scientific Perspectivism*; The University of Chicago Press: Chicago, IL, USA, 2006.
22. Rutishauser, R. Evolution of unusual morphologies in Lentibulariaceae (bladderworts and allies) and Podostemaceae (river-weeds). *Ann. Bot.* **2016**, *117*, 811–832. [CrossRef] [PubMed]
23. Hassenstein, B. Wie viele Körner ergeben einen Haufen? Bemerkungen zu einem uralten und zugleich aktuellen Verständigungsproblem. In *Der Mensch und Seine Sprache*; Peisl, A., Mohler, A., Eds.; Propyläen: Berlin, Germany, 1978; pp. 219–242.
24. Minelli, A. Biological individuality—A complex pattern of distributed uniqueness. In *The Extended Theory of Cognitive Creativity. Perspectives in Pragmatics, Philosophy & Psychology*; Pennisi, A., Falzone, A., Eds.; Springer: Cham, Switzerland, 2020; pp. 185–197.
25. Pradeu, T. Organisms or biological individuals? Combining physiological and evolutionary individuality. *Biol. Philos.* **2016**, *31*, 797–817. [CrossRef]
26. White, J. The plant as a metapopulation. *Annu. Rev. Ecol. Syst.* **1979**, *10*, 109–145. [CrossRef]
27. White, J. Plant metamerism. In *Perspectives on Plant Population Biology*; Dirzo, R., Sarukhan, J., Eds.; Sinauer: Sunderland, MA, USA, 1984; pp. 15–47.
28. Arber, A. *The Natural Philosophy of Plant Form*; Cambridge University Press: Cambridge, UK, 1950.
29. Arber, A. *The Mind and the Eye*; (1964 paperbound reissue); Cambridge University Press: Cambridge, UK, 1954.
30. Arber, A. *The Manifold and the One*; John Murray: London, UK, 1957.
31. Woodger, J.H. *Biological Principles*; Reissued (with new introduction); Humanities: New York, NY, USA, 1967.
32. Rutishauser, R.; Sattler, R. Complementarity and heuristic value of contrasting models in structural botany. I. General considerations. *Bot. Jahrb. Syst.* **1985**, *107*, 415–455.
33. Seibt, J. Ontological tools for the process turn in biology: Some basic notions of general process theory. In *Everything Flows: Towards a Processual Philosophy of Biology*; Dupré, J., Nicholson, D., Eds.; Oxford University Press: Oxford, UK, 2018; pp. 113–136.
34. Kirchoff, B.K.; Pfeifer, E.; Rutishauser, R. Plant structure ontology: How should we label plant structures with doubtful or mixed identities? *Zootaxa* **2008**, *1950*, 103–122. [CrossRef]
35. Rutishauser, R.; Grob, V.; Pfeifer, E. Plants are used to having identity crises. In *Evolving Pathways. Key Themes in Evolutionary Developmental Biology*; Minelli, A., Fusco, G., Eds.; Cambridge Univ. Press: Cambridge, UK, 2008; pp. 194–213.
36. Minelli, A. Morphological misfits and the architecture of development. In *Macroevolution. Explanation, Interpretation and Evidence*; Serrelli, E., Gontier, N., Eds.; Springer: Cham, Switzerland, 2015; pp. 329–343.
37. Minelli, A. Grand challenges in evolutionary developmental biology. *Front. Ecol. Evol.* **2015**, *2*, 85. [CrossRef]
38. Rescher, N. *Process Metaphysics—An Introduction to Process Philosophy*; State University of New York Press: Albany, NY, USA, 1996.
39. Weber, M. (Ed.) *After Whitehead. Rescher on Process Metaphysics*; Ontos Verlag: Frankfurt, Germany; Lancaster, PA, USA, 2004.
40. Nicholson, D.J.; Dupré, J. (Eds.) *Everything Flows: Towards a Processual Philosophy of Biology*; Oxford University Press: Oxford, UK, 2018.
41. Baedke, J.; Mc Manus, S.F. From seconds to eons: Time scales, hierarchies, and processes in evo-devo. *Stud. Hist. Philos. Biol. Biomed. Sci.* **2018**, *72*, 38–48. [CrossRef]
42. Dupré, J. *Processes of Life: Essays in the Philosophy of Biology*; Oxford University Press: Oxford, UK, 2014.

43. Dupré, J.; Guttinger, S. Viruses as living processes. *Stud. Hist. Philos. Biol. Biomed. Sci.* **2016**, *59*, 109–116. [CrossRef]

44. Whitehead, A.N. *Process and Reality: An Essay in Cosmology*; Griffin, D.R., Sherbourne, D.W., Eds.; Macmillan: New York, NY, USA, 1929.

45. Whitehead, A.N. *Modes of Thought*; Macmillan: New York, NY, USA, 1938.

46. Mayr, E. *The Growth of Biological Thought. Diversity, Evolution and Inheritance*; German Edition 1984; Belknap Press of Harvard University Press: Cambridge, MA, USA, 1982; ISBN 0674364457.

47. Jahn, I. (Ed.) *Geschichte der Biologie*; Nikol: Hamburg, Germany, 2000.

48. Mabberley, D.J.; Hay, A. Homoeosis, canalization, decanalization, 'characters' and angiosperm origins. *Edinb. J. Bot.* **1994**, *51*, 117–126. [CrossRef]

49. Lacroix, C.; Jeune, B.; Barabé, D. Encasement in plant morphology: An integrative approach from genes to organisms. *Can. J. Bot.* **2005**, *83*, 1207–1221. [CrossRef]

50. Baedke, J. *Above the Gene, Beyond Biology: Towards a Philosophy of Epigenetics*; University of Pittsburgh Press: Pittsburgh, PA, USA, 2018; ISBN 9780822983408.

51. Benítez, M.; Hernández-Hernández, V.; Newman, S.A.; Niklas, K.J. Dynamical patterning modules, biogeneric materials, and the evolution of multicellular plants. *Front. Plant Sci.* **2018**, *9*, 871. [CrossRef] [PubMed]

52. Wardlaw, C.W. *Organization and Evolution in Plants*; Longmans, Green & Co.: London, UK, 1965.

53. Zimmermann, W. *Die Phylogenie der Pflanzen*, 2nd ed.; G. Fischer: Stuttgart, Germany, 1959.

54. Raff, R.A.; Kaufman, T.C. *Embryos, Genes, and Evolution*; Macmillan: New York, NY, USA, 1983; ISBN 0253206421.

55. Hall, B.K. *Evolutionary Developmental Biology*; Chapman & Hall: London, UK, 1992. [CrossRef]

56. Minelli, A. *Plant Evolutionary Developmental Biology. The Evolvability of the Phenotype*; Cambridge University Press: New York, NY, USA, 2018.

57. Langdale, J.A.; Harrison, C.J. Developmental transitions during the evolution of plant form. In *Evolving Pathways. Key Themes in Evolutionary Developmental Biology*; Minelli, A., Fusco, G., Eds.; Cambridge University Press: Cambridge, UK, 2008; pp. 299–315.

58. Wagner, A. *Arrival of the Fittest: Solving Evolution's Greatest Puzzle*; Penguin: London, UK, 2014.

59. Harrison, J.C. Development and genetics in the evolution of land plant body plans. *Philos. Trans. R. Soc. B* **2017**, *372*, 20150490. [CrossRef]

60. Harrison, C.J.; Morris, J.L. The origin and early evolution of vascular plant shoots and leaves. *Philos. Trans. R. Soc. B Biol. Sci.* **2018**, *373*, 20160496. [CrossRef] [PubMed]

61. Rutishauser, R. Von Goethes dynamischer Pflanzenmorphologie zur evolutionären Entwicklungsbiologie ("EVO-DEVO"): Holismus and Reduktionismus ergänzen sich. *Elem. Naturwiss.* **2018**, *108*, 80–100.

62. Müller, G.B. Homology: The evolution of morphological organization. In *Origination of Organismal Form: Beyond the Gene in Developmental and Evolutionary Biology*; Müller, G.B., Newman, S.A., Eds.; MIT Press: Cambridge, MA, USA, 2003; pp. 51–69.

63. Kauffman, S. *The Origins of Order: Self-Organizing and Selection in Evolution*; Oxford University Press: Oxford, UK, 1993; ISBN 9780195058116.

64. Callebaut, W. Self-organization and optimization: Conflicting or complementary approaches? In *Evolutionary Systems*; Van de Vijver, G., Salthe, S.N., Delpos, M., Eds.; Kluwer: Dordrecht, The Netherlands, 1998; pp. 79–100.

65. Troll, W. *Vergleichende Morphologie der Höheren Pflanzen*; Borntraeger: Berlin, Germany, 1937/1939/1941; Volumes 1/1–3.

66. Baum, D.A. Plant parts: Processes, structures, or functions? *Gard. Bull. Singap.* **2019**, *71* (Suppl. 2), 225–256. [CrossRef]

67. Kaplan, D.R. The science of plant morphology: Definition, history, and role in modern biology. *Am. J. Bot.* **2001**, *88*, 1711–1741. [CrossRef] [PubMed]

68. Kaplan, D.R. Fundamental concepts of leaf morphology and morphogenesis: A contribution to the interpretation of developmental mutants. *Int. J. Plant Sci.* **2001**, *162*, 465–474. [CrossRef]

69. Kinoshita, A.; Tsukaya, H. One-leaf plants in the Gesneriaceae: Natural mutants of the typical shoot system. *Dev. Growth Differ.* **2019**, *61*, 25–33. [CrossRef] [PubMed]

70. Dengler, N.G.; Tsukaya, H. Leaf morphogenesis in dicotyledons: Current issues. *Int. J. Plant Sci.* **2001**, *162*, 459–464. [CrossRef]

71. Frangedakis, E.; Saint-Marcoux, D.; Moody, L.A.; Rabbinowitsch, E.; Langdale, J.A. Nonreciprocal complementation of *KNOX* gene function in land plants. *New Phytol.* **2017**, *216*, 591–604. [CrossRef]

72. Cruz, R.; Melo-de-Pinna, G.F.A.; Vasco, A.; Prado, J.; Ambrose, B.A. Class I KNOX is related to determinacy during the leaf development of the fern *Mickelia scandens* (Dryopteridaceae). *Int. J. Mol. Sci.* **2020**, *21*, 4295. [CrossRef]

73. Ilic, K.; Kellogg, E.A.; Jaiswal, P.; Zapata, F.; Stevens, P.F.; Vincent, L.P.; Avraham, S.; Reiser, L.; Pujar, A.; Sachs, M.M.; et al. The Plant Structure Ontology, a unified vocabulary of anatomy and morphology of a flowering plant. *Plant Physiol.* **2007**, *143*, 587–599. [CrossRef]

74. Schneider, H. Evolutionary morphology of ferns (Monilophytes). *Annu. Plant Rev.* **2013**, *45*, 115–140. [CrossRef]

75. Bell, A.D. *Plant Form—An Illustrated Guide to Flowering Plant Morphology*; Oxford University Press: Oxford, UK, 1991.

76. Theissen, G. The proper place of hopeful monsters in evolutionary biology. *Theory Biosci.* **2006**, *124*, 349–369. [CrossRef]

77. Theissen, G. Saltational evolution: Hopeful monsters are here to stay. *Theory Biosci.* **2009**, *128*, 43–51. [CrossRef]

78. Masel, J.; Siegal, M.L. Robustness: Mechanisms and consequences. *Trends Genet.* **2009**, *25*, 395–403. [CrossRef] [PubMed]

79. Arthur, W. *Evolution: A Developmental Approach*; Wiley Blackwell: Chichester, UK, 2011.

80. Minelli, A. Tracing homologies in an ever-changing world. *Riv. Estet.* **2016**, *62*, 40–55. [CrossRef]

81. Lemon, G.; Posluszny, U. Comparative shoot development and evolution in the Lemnaceae. *Int. J. Plant Sci.* **2000**, *161*, 733–748. [CrossRef]

82. Katayama, N.; Koi, S.; Kato, M. Expression of *Shoot Meristemless, Wuschel*, and *Asymmetric Leaves1* homologs in the shoots of Podostemaceae: Implications for the evolution of novel shoot organogenesis. *Plant Cell* **2010**, *22*, 2131–2140. [CrossRef] [PubMed]

83. Katayama, N.; Kato, M.; Yamada, T. Origin and development of the cryptic shoot meristem in *Zeylanidium lichenoides*. *Am. J. Bot.* **2013**, *100*, 635–646. [CrossRef]

84. Kato, M. *The Illustrated Book of Plant Systematics in Color: Podostemaceae of the World*; (In Japanese, with English Summaries); Hokuryukan: Tokyo, Japan, 2013.

85. Tsukaya, H. Comparative leaf development in angiosperms. *Curr. Opin. Plant Biol.* **2014**, *17*, 103–109. [CrossRef]

86. Arber, A. Root and shoot in the angiosperms: A study of morphological categories. *New Phytol.* **1930**, *29*, 297–315. [CrossRef]

87. Howard, R.A. The stem-node-leaf continuum of the Dicotyledoneae. *J. Arnold Arbor.* **1974**, *55*, 125–181.

88. Cusset, G. The conceptual bases of plant morphology. *Acta Biotheor.* **1982**, *31*, 8–86.

89. Cusset, G. A simple classification of the complex parts of vascular plants. *Bot. J. Linn. Soc.* **1994**, *114*, 229–242. [CrossRef]

90. Sattler, R. (Ed.) *Axioms and Principles of Plant Construction*; (Acta Biotheoretica 31A); Nijhoff/Junk: The Hague, The Netherlands, 1982.

91. Warming, E. Om forskjellen mellem trichomer og epiblastemer af höjere rang. *Vidensk. Medd. Dansk. Naturhist. Foren. Kjobenhavn* **1872**, *16–27*, 159–205.

92. von Goethe, J.W. *Versuch die Metamorphose der Pflanzen zu Erklären*; Ettingersche Buchhandlung: Gotha, Germany, 1790.

93. Müller-Xing, R.; Schubert, D.; Goodrich, J. Non-inductive conditions expose the cryptic bract of flower phytomeres in *Arabidopsis thaliana*. *Plant Signal. Behav.* **2015**, *10*, e1010868. [CrossRef]

94. Lyndon, R.F. The mechanism of leaf initiation. In *The Growth and Functioning of Leaves*; Dale, J.E., Milthorpe, F.L., Eds.; Cambridge University Press: Cambridge, UK, 1983; pp. 3–24.

95. Dickinson, T.A. Epiphylly in angiosperms. *Bot. Rev.* **1978**, *44*, 181–232. [CrossRef]

96. Garcês, H.M.P.; Champagne, C.E.M.; Townsley, B.T.; Park, S.; Malhó, R.; Pedroso, M.C.; Harada, J.J.; Sinha, N.R. Evolution of asexual reproduction in leaves of the genus *Kalanchoë*. *Proc. Natl. Acad. Sci. USA* **2007**, *104*, 15578–15583. [CrossRef] [PubMed]

97. Saunders, E.R. The leaf-skin theory of the stem: A consideration of certain anatomico-physiological relations in the spermatophyte shoot. *Ann. Bot.* **1922**, *36*, 135–165. [CrossRef]

98. Evans, M.W.; Grover, F.O. Developmental morphology of the growing point of the shoot and the inflorescence in grasses. *J. Agric. Res.* **1940**, *61*, 481–520.

99. Rohweder, O. Anatomische und histogenetische Untersuchungen an Laubsprossen und Blüten der Commelinaceen. *Bot. Jahrb. Syst.* **1963**, *82*, 1–99.

100. Vita, R.S.B.; Menezes, N.L.; Pellegrini, M.O.O.; Melo-de-Pinna, G.F.A. A new interpretation on vascular architecture of the cauline system in Commelinaceae (Commelinales). *PLoS ONE* **2019**, *14*, e0218383. [CrossRef]

101. Sattler, R. Ein neues Spross-Modell. *Ber. Dtsch. Bot. Ges.* **1971**, *84*, 139.

102. Sattler, R. A new conception of the shoot of higher plants. *J. Theor. Biol.* **1974**, *47*, 367–382. [CrossRef]

103. Arber, A. The interpretation of leaf and root in the angiosperms. *Biol. Rev.* **1941**, *16*, 81–105. [CrossRef]

104. Cruz, R.; Prado, J.; Melo-de-Pinna, G.F.A. Leaf development in some ferns with variable dissection patterns (Dryopteridaceae and Lomariopsidaceae). *Flora* **2020**, 151658. [CrossRef]

105. Mueller, B. (Ed.) *Goethe's Botanical Writings*; Ox Bow Press: Woodbridge, CT, USA, 1989.

106. Minelli, A. Limbs and tail as evolutionarily diverging duplicates of the main body axis. *Evol. Dev.* **2000**, *2*, 157–165. [CrossRef] [PubMed]

107. Minelli, A. *The Development of Animal Form: Ontogeny, Morphology, and Evolution*; Cambridge University Press: Cambridge, UK, 2003.

108. Rutishauser, R.; Moline, P. Evo-devo and the search for homology ('sameness') in biological systems. *Theory Biosci.* **2005**, *124*, 213–241. [CrossRef]

109. Plackett, A.R.G.; Di Stilio, V.S.; Langdale, J.A. Ferns: The missing link in shoot evolution and development. *Front. Plant Sci.* **2015**, *6*, 972. [CrossRef] [PubMed]

110. Beerling, D.J. Leaf evolution: Gases, genes and geochemistry. *Ann. Bot.* **2005**, *96*, 345–352. [CrossRef] [PubMed]

111. Feild, T.S.; Brodribb, T.J.; Iglesias, A.; Chatelet, D.S.; Baresch, A.; Upchurch, G.R., Jr.; Gomez, B.; Mohr, B.A.R.; Coiffard, C.; Kvacek, J.; et al. Fossil evidence for Cretaceous escalation in angiosperm leaf vein evolution. *Proc. Natl. Acad. Sci. USA* **2011**, *108*, 8363–8366. [CrossRef]

112. Huxley, J.S. *Introduction to Vertebrate Zoology*; de Beer, G.R., Ed.; Macmillan Co.: New York, NY, USA, 1928.

113. Sattler, R. Homology—A continuing challenge. *Syst. Bot.* **1984**, *9*, 382–394. [CrossRef]

114. Minelli, A.; Fusco, G. Homology. In *The Philosophy of Biology: A Companion for Educators*; Kampourakis, K., Ed.; Springer: Dordrecht, The Netherlands, 2013; pp. 289–322. [CrossRef]

115. Ochoterena, H.; Vrijdaghs, A.; Smets, E.; Classen-Bockhoff, R. The search for common origin: Homology revisited. *Syst. Biol.* **2019**, *68*, 767–780. [CrossRef]

116. Owen, R. *Lectures on Comparative Anatomy and Physiology of the Invertebrate Animals, Delivered at the Royal College of Surgeons in 1843*; Longman, Brown, Green & Longmans: London, UK, 1843.

117. Remane, A. *Die Grundlagen des Natürlichen Systems, der Vergleichenden Anatomie und der Phylogenetik*, 2nd ed.; Geest et Portig: Leipzig, Germany, 1956.

118. Eckardt, T.H. Das Homologieproblem und Fälle strittiger Homologien. *Phytomorphology* **1964**, *14*, 79–92.

119. Sattler, R. Homology, homeosis, and process morphology in plants. In *Homology: The Hierarchical Basis of Comparative Morphology*; Hall, B.K., Ed.; Academic Press: New York, NY, USA, 1994; pp. 423–475.

120. Brigandt, I. Homology in comparative, molecular and evolutionary developmental biology: The radiation of a concept. *J. Exp. Zool.* **2003**, *299*, 9–17. [CrossRef] [PubMed]

121. Abouheif, E. Developmental genetics and homology: A hierarchical approach. *Trends Ecol. Evol.* **1997**, *12*, 405–408. [CrossRef]

122. Abouheif, E. Establishing homology criteria for regulatory gene networks: Prospects and challenges. In *Homology*; Bock, G.R., Cardew, G., Eds.; Wiley: Chichester, UK, 1999; pp. 207–225.

123. Brigandt, I. How are biology concepts used and transformed? In *Philosophy of Science for Biologists*; Kampourakis, K., Uller, T., Eds.; University Press Cambridge: Cambridge, UK, 2020; pp. 79–101.

124. Jaramillo, M.A.; Kramer, E.M. The role of developmental genetics in understanding homology and morphological evolution in plants. *Int. J. Plant Sci.* **2007**, *168*, 61–72. [CrossRef]

125. Koentges, G. Evolution of anatomy and gene control. Evo–devo meets systems biology. *Nature* **2008**, *451*, 658–663. [CrossRef]

126. Shubin, N.; Tabin, C.; Carroll, S. Fossils, genes and the evolution of animal limbs. *Nature* **1997**, *388*, 639–648. [CrossRef]

127. Wilkins, A.S. *The Evolution of Developmental Pathways*; Sinauer: Sunderland, MA, USA, 2002.

128. Blochlinger, K.; Jan, L.Y.; Jan, Y.N. Transformation of sensory organ identity by ectopic expression of Cut in *Drosophila*. *Genes Dev.* **1991**, *5*, 1124–1135. [CrossRef]

129. Moreau, C.; Hofer, J.M.I.; Eléouet, M.; Sinjushin, A.; Ambrose, M.; Skøt, K.; Blackmore, T.; Swain, M.; Hegarty, M.; Balanzà, V.; et al. Identification of *Stipules reduced*, a leaf morphology gene in pea (*Pisum sativum*). *New Phytol.* **2018**, *220*, 288–299. [CrossRef]

130. Sylvester, A.W.; Smith, L.; Freeling, M. Acquisition of identity in the developing leaf. *Annu. Rev. Cell Dev. Biol.* **1996**, *12*, 257–304. [CrossRef]

131. Steeves, T.A.; Hicks, G.; Steeves, M.; Retallack, B. Leaf determination in the fern *Osmunda cinnamomea*: A reinvestigation. *Ann. Bot.* **1993**, *71*, 511–517. [CrossRef]

132. Claßen-Bockhoff, R.; Frankenhäuser, H. The 'Male Flower' of *Ricinus communis* (Euphorbiaceae) interpreted as a multi-flowered unit. *Front. Cell Dev. Biol.* **2020**, *8*, 313. [CrossRef]

133. Sattler, R. Towards a more adequate approach to comparative morphology. *Phytomorphology* **1966**, *16*, 417–429.

134. Sattler, R. Philosophy of plant morphology. *Elem. Nat.* **2018**, *108*, 55–79.

135. Sattler, R. Process morphology: Structural dynamics in development and evolution. *Can. J. Bot.* **1992**, *70*, 708–714. [CrossRef]

136. Langdale, J.A. Evolution of developmental mechanisms in plants. *Curr. Opin. Gen. Dev.* **2008**, *18*, 368–373. [CrossRef] [PubMed]

137. Vergara-Silva, F. Plants and the conceptual articulation of evolutionary developmental biology. *Biol. Philos.* **2003**, *18*, 261–264. [CrossRef]

138. Jeune, B.; Barabe, D.; Lacroix, C. Classical and dynamic morphology: Toward a synthesis through the space of forms. *Acta Biotheor.* **2006**, *54*, 277–293. [CrossRef] [PubMed]

139. Lacroix, C.; Jeune, B.; Purcell-MacDonald, S. Shoot and compound leaf comparisons in eudicots: Dynamic morphology as an alternative approach. *Bot. J. Linn. Soc.* **2003**, *143*, 219–230. [CrossRef]

140. Niklas, K.J. *The Evolutionary Biology of Plants*; The University of Chicago Press: Chicago, IL, USA, 1997.

141. Cronk, Q. *The Molecular Organography of Plants*; Oxford University Press: Oxford, UK, 2009.

142. Donoghue, M.J.; Kadereit, J.W. Walter Zimmermann and the growth of phylogenetic theory. *Syst. Biol.* **1992**, *41*, 74–85. [CrossRef]

143. Zimmermann, W. Main results of the 'Telome Theory'. *Palaeobotanist* **1952**, *1*, 456–470.

144. Floyd, S.K.; Bowman, J.L. The ancestral developmental tool kit of land plants. *Int. J. Plant Sci.* **2007**, *168*, 1–35. [CrossRef]

145. Beerling, D.J.; Fleming, A.J. Zimmermann's telome theory of megaphyll leaf evolution. *Curr. Opin. Plant Biol.* **2007**, *10*, 4–12. [CrossRef] [PubMed]

146. Tomescu, A.M.F. Megaphylls, microphylls and the evolution of leaf development. *Trends Plant Sci.* **2009**, *14*, 5–12. [CrossRef] [PubMed]

147. Boyce, C.K. The evolution of plant development in a paleontological context. *Curr. Opin. Plant Biol.* **2010**, *13*, 102–107. [CrossRef] [PubMed]

148. Harrison, C.J.; Coriey, S.B.; Moylan, E.C.; Alexander, D.L.; Scotland, R.W.; Langdale, J.A. Independent recruitment of a conserved developmental mechanism during leaf evolution. *Nature* **2005**, *434*, 509–514. [CrossRef] [PubMed]

149. Gilbert, S.F.; Bolker, J.A. Homologies of process and modular elements of embryonic construction. *J. Exp. Zool.* **2001**, *291*, 1–12. [CrossRef]

150. Albert, V.A.; Jobson, R.W. *Relaxed Structural Constraints in Utricularia (Lentibulariaceae): A Possible Basis in One or Few Genes Regulating Polar Auxin Transport*; Abstract, AIBS Meeting Albuquerque: New Mexico, NM, USA, 2001.

151. Sinha, N.R. Leaf development in angiosperms. *Ann. Rev. Plant Physiol. Plant Mol. Biol.* **1999**, *50*, 419–446. [CrossRef]

152. Hofer, J.M.I.; Gourlay, C.W.; Ellis, T.H.N. Genetic control of leaf morphology: A partial view. *Ann. Bot.* **2001**, *88*, 1129–1139. [CrossRef]

153. Baum, D.A.; Donoghue, M.J. Transference of function, heterotopy and the evolution of plant development. In *Developmental Genetics and Plant Evolution*; Cronk, Q.C.B., Bateman, R.M., Hawkins, J.A., Eds.; Taylor & Francis: London, UK, 2002; pp. 52–69.

154. James, P.J. Tree and Leaf: A different angle. *Linnean* **2009**, *25*, 13–19.

155. Prusinkiewicz, P.; Runions, A. Computational models of plant development and form. *New Phytol.* **2012**, *193*, 549–569. [CrossRef]

156. Jeune, B.; Sattler, R. Multivariate analysis in process morphology. *J. Theor. Biol.* **1992**, *156*, 147–167. [CrossRef]

157. Sattler, R.; Jeune, B. Multivariate analysis confirms the continuum view of plant form. *Ann. Bot.* **1992**, *69*, 249–262. [CrossRef]

158. Tautz, D. The continued mystery of the phylotypic stage. In *Perspectives on Evolutionary and Developmental Biology: Essays for Alessandro Minelli*; Fusco, G., Ed.; Padova University Press: Padova, Italy, 2019; pp. 141–149.

159. Gilbert, S.F.; Bard, J. Formalizing theories of development: A fugue on the orderliness of change. In *Towards a Theory of Development*; Minelli, A., Pradeu, T., Eds.; Oxford University Press: Oxford, UK; New York, NY, USA, 2014; pp. 129–143. [CrossRef]

160. Wolpert, L.; Beddington, R.; Jessell, T.; Lawrence, P.; Meyerowitz, E.; Smith, J. *Principles of Development*, 2nd ed.; Oxford University Press: Oxford, UK, 2002.

161. Butler, A.; Saidel, W.M. Defining sameness: Historical, biological, and generative homology. *BioEssays.* **2000**, *22*, 846–853. [CrossRef]

162. DiFrisco, J. Developmental Homology. In *Evolutionary Developmental Biology: A Reference Guide*; Nuno de la Rosa, L., Müller, G., Eds.; Springer: Cham, Switzerland, 2021.

163. Kessler, S.; Sinha, N. Shaping up: The genetic control of leaf shape. *Curr. Opin. Plant Biol.* **2004**, *7*, 65–72. [CrossRef]

164. Kim, M.; Pham, T.; Hamidi, A.; McCormick, S.; Kuzoff, R.K.; Sinha, N. Reduced leaf complexity in tomato wiry mutants suggests a role for PHAN and KNOX genes in generating compound leaves. *Development* **2003**, *130*, 4405–4415. [CrossRef]

165. Champagne, C.E.M.; Goliber, T.E.; Wojciechowski, M.F.; Mei, R.W.; Townsley, B.T.; Wang, K.; Paz, M.M.; Geeta, R.; Sinha, N.R. Compound leaf development and evolution in the Legumes. *Plant Cell.* **2007**, *19*, 3369–3378. [CrossRef] [PubMed]

166. Weiss, K.M. The phenogenetic logic of life. *Nat. Rev. Genet.* **2005**, *6*, 36–46. [CrossRef]

167. Sussex, I.M.; Kerk, N.M. The evolution of plant architecture. *Curr. Opin. Plant Biol.* **2001**, *4*, 33–37. [CrossRef]

168. Prusinkiewicz, P.; Lindenmayer, A. *The Algorithmic Beauty of Plants*; Springer: New York, NY, USA, 1990.

169. Tsukaya, H. The role of meristematic activities in the formation of leaf blades. *J. Plant Res.* **2000**, *113*, 119–126. [CrossRef]

170. Yamaguchi, T.; Nukazuka, A.; Tsukaya, H. Leaf adaxial–abaxial polarity specification and lamina outgrowth: Evolution and development. *Plant Cell Physiol.* **2012**, *53*, 1180–1194. [CrossRef]

171. Nakayama, H.; Yamaguchi, T.; Tsukaya, H. Acquisition and diversification of cladodes: Leaf-like organs in the genus *Asparagus*. *Plant Cell* **2012**, *24*, 929–940. [CrossRef]

172. Cooney-Sovetts, C.; Sattler, R. Phylloclade development in the Asparagaceae: An example of homeosis. *Bot. J. Linn. Soc.* **1987**, *94*, 327–371. [CrossRef]

173. Stewart, W.N.; Rothwell, G.W. *Paleobotany and the Evolution of Plants*; Cambridge University Press: Cambridge, UK, 1993.

174. Sanders, H.L.; Darrah, P.R.; Langdale, J.A. Sector analysis and predictive modelling reveal iterative shoot-like development in fern fronds. *Development* **2011**, *138*, 2925–2934. [CrossRef] [PubMed]

175. Hébant-Mauri, R. The branching of *Trichomanes proliferum* (Hymenophyllaceae). *Can. J. Bot.* **1990**, *68*, 1091–1097. [CrossRef]

176. Steeves, T.A.; Sussex, I.M. *Patterns in Plant Development*, 2nd ed.; Cambridge University Press: Cambridge, UK, 1989.

177. Vasco, A.; Ambrose, B.A. Simple and divided leaves in ferns: Exploring the genetic basis for leaf morphology differences in the genus *Elaphoglossum* (Dryopteridaceae). *Int. J. Mol. Sci.* **2020**, *21*, 5180. [CrossRef] [PubMed]

178. Richards, J.H.; Beck, J.Z.; Hirsch, A.M. Structural investigations of asexual reproduction in *Nephrolepis exaltata* and *Platycerium bifurcatum*. *Am. J. Bot.* **1983**, *70*, 993–1001. [CrossRef]

179. Reiser, L.; Sanchez-Baracaldo, P.; Hake, S. Knots in the family tree: Evolutionary relationships and functions of KNOX homeobox genes. *Plant Mol. Biol.* **2000**, *42*, 151–166. [CrossRef]

180. Vasco, A.; Moran, R.C.; Ambrose, B.A. The evolution, morphology, and development of fern leaves. *Front. Plant Sci.* **2013**, *4*, 345. [CrossRef]

181. Champagne, C.E.M.; Sinha, N. Compound leaves: Equal to the sum of their parts? *Development* **2004**, *131*, 4401–4412. [CrossRef]

182. Smith, L.G.; Hake, S. The initiation and determination of leaves. *Plant Cell* **1992**, *4*, 1017–1027. [CrossRef]

183. Sattler, R.; Rutishauser, R. Partial homology of pinnate leaves and shoots: Orientation of leaflet inception. *Bot. Jahrb. Syst.* **1992**, *114*, 61–79.

184. Lacroix, C.R.; Sattler, R. Expression of shoot features in early leaf development of *Murraya paniculata* (Rutaceae). *Can. J. Bot.* **1994**, *72*, 678–687. [CrossRef]

185. Lacroix, C.R. Changes in leaflet and leaf lobe form in developing compound and finely divided leaves. *Bot. Jahrb. Syst.* **1995**, *117*, 317–331.

186. Rutishauser, R. Developmental patterns of leaves in Podostemaceae compared with more typical flowering plants: Saltational evolution and fuzzy morphology. *Can. J. Bot.* **1995**, *73*, 1305–1317. [CrossRef]

187. Jeune, B.; Lacroix, C.R. A quantitative model of leaflet initiation illustrated by *Murraya paniculata* (Rutaceae). *Can. J. Bot.* **1993**, *71*, 457–465. [CrossRef]

188. Bharathan, G.; Sinha, N.R. The regulation of compound leaf development. *Plant Physiol.* **2001**, *127*, 1533–1538. [CrossRef]

189. Kaplan, D.R. Comparative developmental evaluation of the morphology of unifacial leaves in the monocotyledons. *Bot. Jahrb. Syst.* **1975**, *95*, 1–105.

190. Hagemann, W. *Morphological Aspects of Leaf Development in Ferns and Angiosperms*; Academic Press: New York, NY, USA, 1984.

191. Fisher, J.B.; Rutishauser, R. Leaves and epiphyllous shoots in *Chisocheton* (Meliaceae), a continuum of woody leaf and stem axes. *Can. J. Bot.* **1990**, *68*, 2316–2328. [CrossRef]

192. Melo-de-Pinna, G.F.A.; Cruz, R. Leaf development in vascular plants. In *Plant Ontogeny*; Demarco, D., Ed.; Nova Science Publ.: Hauppauge, NY, USA, 2020; pp. 83–105.

193. Bar, M.S.; Ori, N. Leaf development and morphogenesis. *Development* **2014**, *141*, 4219–4230. [CrossRef]

194. Bar, M.S.; Ori, N. Compound leaf development in model plant species. *Curr. Opin. Plant Biol.* **2015**, *23*, 61–69. [CrossRef] [PubMed]

195. Steingraeber, D.A.; Fisher, J.B. Indeterminate growth of leaves in *Guarea* (Meliaceae): A twig analogue. *Am. J. Bot.* **1986**, *73*, 852–862. [CrossRef]

196. Fukuda, T.; Yokoyama, J.; Tsukaya, H. Phylogenetic relationships among species in the genera *Chisocheton* and *Guarea* that have unique indeterminate leaves as inferred from sequences of chloroplast DNA. *Int. J. Plant Sci.* **2001**, *164*, 13–24. [CrossRef]

197. Fisher, J.B. Indeterminate leaves of *Chisocheton* (Meliaceae): Survey of structure and development. *Bot. J. Linn. Soc.* **2002**, *139*, 207–221. [CrossRef]

198. Stevens, P.F. Review of *Chisocheton* (Meliaceae) in Papuasia. *Contrib. Herb. Aust.* **1975**, *11*, 1–55.

199. Rutishauser, R. Polymerous leaf whorls in vascular plants: Developmental morphology and fuzziness of organ identities. *Int. J. Plant Sci.* **1999**, *160* (Suppl. 6), S81–S103. [CrossRef] [PubMed]

200. Rutishauser, R.; Sattler, R. Architecture and development of the phyllode-stipules whorls in *Acacia longipedunculata*: Controversial interpretations and continuum approach. *Can. J. Bot.* **1986**, *64*, 1987–2019. [CrossRef]

201. Cruz, R.; Duarte, M.; Pirani, J.R.; Melo-dePinna, G.F.A. Development of leaves and shoot apex protection in *Metrodorea* and related species (Rutaceae). *Bot. J. Linn. Soc.* **2015**, *178*, 267–282. [CrossRef]

202. Yaxley, J.L.; Jablonski, W.; Reid, J.B. Leaf and flower development in pea (*Pisum sativum* L.): Mutants *cochleata* and *unifoliata*. *Ann. Bot.* **2001**, *88*, 225–234. [CrossRef]

203. Marx, G.A. A suite of mutants that modify pattern formation in pea leaves. *Plant Mol. Biol. Report.* **1987**, *5*, 311–335. [CrossRef]

204. Tattersall, A.D.; Turner, L.; Knox, M.R.; Ambrose, M.J.; Ellis, T.H.N.; Hofer, J.M.I. The mutant *crispa* reveals multiple roles for PHANTASTICA in pea compound leaf development. *Plant Cell* **2005**, *17*, 1046–1060. [CrossRef]

205. Jong, K.; Burtt, B.L. The evolution of morphological novelty exemplified in the growth patterns of some Gesneriaceae. *New Phytol.* **1975**, *75*, 297–311. [CrossRef]

206. Tsukaya, H. Determination of the unequal fate of cotyledons of a one-leaf plant, *Monophyllaea*. *Development* **1997**, *124*, 1275–1280. [PubMed]

207. Weber, A.; Clark, J.L.; Möller, M. A new formal classification of Gesneriaceae. *Selbyana* **2013**, *31*, 68–94.

208. Kinoshita, A.; Koga, H.; Tsukaya, H. Expression profiles of *ANGUSTIFOLIA3* and *SHOOT MERISTEMLESS*, key genes for meristematic activity in a one-leaf plant *Monophyllaea glabra*, revealed by whole-mount in situ hybridization. *Front. Plant Sci.* **2020**, *11*, 1160. [CrossRef]

209. Nishii, K.; Spada, A.; Möller, M. Hormonal crosstalk in the regulation of meristem activity and the phyllomorph architecture in *Streptocarpus* (Gesneriaceae): A review. *Rheedea* **2020**, *30*, 96. [CrossRef]

210. Eckardt, N.A.; Baum, D. The podostemad puzzle: The evolution of unusual morphology in the Podostemaceae. *Plant Cell* **2010**, *22*, 2131–2140. [CrossRef] [PubMed]

211. Arber, A. *Water Plants. A Study of Aquatic Angiosperms*; Cambridge University Press: Cambridge, UK, 1920.

212. Taylor, P. *The Genus Utricularia—A Taxonomic Monograph*; HMSO: London, UK, 1989.

213. Jobson, R.W.; Baleeiro, P.C.; Guisande, C. Systematics and evolution of Lentibulariaceae: III. *Utricularia*. In *Carnivorous Plants: Physiology, Ecology, and Evolution*; Ellison, A.M., Adamec, L., Eds.; Oxford Univ. Press: Oxford, UK, 2018; pp. 89–104.

214. Silva, S.R.; Moraes, A.P.; Penha, H.A.; Julião, M.H.M.; Domingues, D.S.; Michael, T.P.; Miranda, V.F.O.; Varani, A.M. The terrestrial carnivorous plant *Utricularia reniformis* sheds light on environmental and life-form genome plasticity. *Int. J. Mol. Sci.* **2020**, *21*, 3. [CrossRef] [PubMed]

215. Whitewoods, C.D. Quick Guide—*Utricularia*. *Curr. Biol.* **2020**, *30*, R143–R144. [CrossRef]

216. Whitewoods, C.D.; Gonçalves, B.; Cheng, J.; Cui, M.; Kennaway, R.; Lee, K.; Bushell, C.; Yu, M.; Piao, C.; Coen, E. Evolution of carnivorous traps from planar leaves through simple shifts in gene expression. *Science* **2020**, *367*, 91–96. [CrossRef]

217. Goebel, K. *Pflanzenbiologische Schilderungen*; Part II; Elwert: Marburg, Germany, 1891.

218. Reut, M.S.; Plachno, B.J. Unusual developmental morphology and anatomy of vegetative organs in *Utricularia dichotoma*—Leaf, shoot and root dynamics. *Protoplasma* **2020**, *257*, 371–390. [CrossRef]

219. Carretero-Paulet, L.; Chang, T.-H.; Librado, P.; Ibarra-Laclette, E.; Herrera-Estrella, L.; Rozas, J.; Albert, V.A. Genome-wide analysis of adaptive molecular evolution in the carnivorous plant *Utricularia gibba*. *Genome Biol. Evol.* **2015**, *7*, 444–456. [CrossRef]

220. Carretero-Paulet, L.; Librado, P.; Chang, T.-H.; Ibarra-Laclette, E.; Herrera-Estrella, L.; Rozas, J.; Albert, V.A. High gene family turnover rates and gene space adaptation in the compact genome of the carnivorous plant *Utricularia gibba*. *Mol. Biol. Evol.* **2015**, *32*, 1284–1295. [CrossRef]

221. Albert, V.A.; Jobson, R.W.; Michael, T.P.; Taylor, D.J. The carnivorous bladderwort (*Utricularia*, Lentibulariaceae): A system inflates. *J. Exp. Bot.* **2010**, *61*, 5–9. [CrossRef]

222. Barta, J.; Stone, J.D.; Pech, J.; Sirová, D.; Adamec, L.; Campbell, M.A.; Štorchová, H. The transcriptome of *Utricularia vulgaris*, a rootless plant with minimalist genome, reveals extreme alternative splicing and only moderate similarity with *Utricularia gibba*. *BMC Plant Biol.* **2015**, *15*, 78. [CrossRef]

223. Renner, T.; Lan, T.; Farr, K.M.; Ibarra-Laclette, E.; Herrera-Estrella, L.; Schuster, S.C.; Hasebe, M.; Fukushima, K.; Albert, V.A. Carnivorous plant genomes. In *Carnivorous Plants: Physiology, Ecology, and Evolution*; Ellison, A.M., Adamec, L., Eds.; Oxford Univ. Press: Oxford, UK, 2018; pp. 135–153.

224. Brugger, J.; Rutishauser, R. Bau und Entwicklung landbewohnender *Utricularia*-Arten. *Bot. Helv.* **1989**, *99*, 91–146.

225. Guédès, M. *Morphology of Seed-Plants*; J. Cramer: Vaduz, Liechtenstein, 1979.

226. Ibarra-Laclette, E.; Albert, V.A.; Perez-Torres, C.A.; Zamudio-Hernández, F.; Ortega-Estrada, M.J.; Herrera-Estrella, A. Transcriptomics and molecular evolutionary rate analysis of the bladderwort (*Utricularia*), a carnivorous plant with a minimal genome. *BMC Plant Biol.* **2011**, *11*, 101. [CrossRef] [PubMed]

227. Ibarra-Laclette, E.; Lyons, E.; Hernandez-Guzman, G.; Pérez-Torres, C.A.; Carretero-Paulet, L.; Chang, T.H.; Lan, T.; Welch, A.J.; Juárez, M.J.A.; Fernández-Cortés, A.; et al. Architecture and evolution of a minute plant genome. *Nature* **2013**, *498*, 94–98. [CrossRef] [PubMed]

228. Cronk, Q.C.B. Plant evolution and development in a post-genomic context. *Nat. Rev. Genet.* **2001**, *2*, 607–619. [CrossRef]

229. Hofhuis, H.; Laskowski, M.; Du, Y.; Prasad, K.; Grigg, S.; Pinon, V.; Scheres, B. Phyllotaxis and rhizotaxis in *Arabidopsis* are modified by three PLETHORA transcription factors. *Curr. Biol.* **2013**, *23*, 956–962. [CrossRef] [PubMed]

230. Friedman, W.E.; Moore, R.C.; Purugganan, M.D. The evolution of plant development. *Am. J. Bot.* **2004**, *91*, 1726–1741. [CrossRef] [PubMed]

231. Raven, J.A.; Edwards, D. Roots: Evolutionary origins and biogeochemical significance. *J. Exp. Bot.* **2001**, *52*, 381–401. [CrossRef]

232. Schneider, H.; Pryer, K.M.; Cranfill, R.; Smith, A.R.; Wolf, P.G. The evolution of vascular plant body plans— A phylogenetic perspective. In *Developmental Genetics and Plant Evolution*; Cronk, Q.C.B., Bateman, R.M., Hawkins, J.A., Eds.; Taylor & Francis: London, UK, 2002; pp. 330–364.

233. Sattler, R. *Organogenesis of Flowers. A Photographic Text-Atlas*; Univ. of Toronto Press: Toronto, ON, Canada, 1973.

234. Endress, P.K. *Diversity and Evolutionary Biology of Tropical Flowers*; Cambridge University Press: Cambridge, UK, 1994.

235. Soltis, D.; Soltis, P.; Endress, P.; Chase, M.; Manchester, S.; Judd, W.; Majure, L.; Mavrodiev, E. *Phylogeny and Evolution of the Angiosperms*; Revised and Updated Edition; The University of Chicago Press: Chicago, IL, USA; London, UK, 2018.

236. Johnson, S.D.; Schiestl, F.P. *Floral Mimicry*; Oxford Univ. Press: Oxford, UK, 2016.

237. De Ronse Craene, L.P. *Floral Diagrams: An Aid to Understanding Flower Morphology and Evolution*; Cambridge University Press: Cambridge, UK, 2010

238. Coen, E.S.; Meyerowitz, E.M. The war of the whorls: Genetic interactions controlling flower development. *Nature* **1991**, *353*, 31–37. [CrossRef]

239. Theissen, G.; Melzer, R. Molecular mechanisms underlying origin and diversification of the angiosperm flower. *Ann. Bot.* **2007**, *100*, 603–619. [CrossRef]

240. Endress, P.K. Angiosperm floral evolution: Morphological and developmental framework. *Adv. Bot. Res.* **2006**, *44*, 1–61.

241. Chanderbali, A.S.; Berger, B.A.; Howarth, D.G.; Soltis, D.E.; Soltis, P.S. Evolution of floral diversity: Genomics, genes and gamma. *Philos. Trans. R. Soc. B* **2017**, *372*, 20150509. [CrossRef] [PubMed]

242. Soltis, D.E.; Chanderbali, A.S.; Kim, S.; Buzgo, M.; Soltis, P.S. The ABC model and its applicability to basal angiosperms. *Ann. Bot.* **2007**, *100*, 155–163. [CrossRef]

243. Soltis, P.S.; Soltis, D.E. Flower Diversity and Angiosperm Diversification. In *Flower Development*; Riechmann, J.L., Wellmer, F., Eds.; Springer: New York, NY, USA; Berlin/Heidelberg, Germany, 2014; pp. 85–102.

244. Buzgo, M.; Soltis, P.S.; Soltis, D.E. Floral developmental morphology of *Amborella trichopoda* (Amborellaceae). *Int. J. Plant Sci.* **2004**, *165*, 925–947. [CrossRef]

245. Buzgo, M.; Soltis, P.S.; Kim, S.; Soltis, D.E. The making of the flower. *Biologist* **2005**, *52*, 149–154.

246. Warner, K.A.; Rudall, P.J.; Frohlich, M.W. Environmental control of sepalness and petalness in perianth organs of waterlilies—A new Mosaic Theory on the evolutionary origin of a differentiated perianth. *J. Exp. Bot.* **2009**, *60*, 3559–3574. [CrossRef] [PubMed]

247. De Candolle, A.P. *Organographie Végétale*; Deterville: Paris, France, 1827; Volume 2.

248. Prenner, G.; Rudall, P.J. Comparative ontogeny of the cyathium in *Euphorbia* (Euphorbiaceae) and its allies: Exploring the organ-flower-inflorescence boundary. *Am. J. Bot.* **2007**, *94*, 1612–1629. [CrossRef]

249. Sokoloff, D.; Rudall, P.J.; Remizowa, M. Flower-like terminal structures in racemose inflorescences: A tool in morphogenetic and evolutionary research. *J. Exp. Bot.* **2006**, *57*, 3517–3530. [CrossRef]

250. Prenner, G.; Box, M.S.; Cunniff, J.; Rudall, P.J. Branching stamens of *Ricinus* and the homologies of the angiosperm stamen fascicle. *Int. J. Plant Sci.* **2008**, *169*, 735–744. [CrossRef]

251. Prenner, G.; Cacho, N.I.; Baum, D.; Rudall, P.J. Is LEAFY a useful marker gene for the flower-inflorescence boundary in the *Euphorbia* cyathium? *J. Exp. Bot.* **2011**, *62*, 345–350. [CrossRef] [PubMed]

252. Vekemans, D.; Viaene, T.; Caris, P.; Geuten, K. Transference of function shapes organ identity in the dove tree inflorescence. *New Phytol.* **2012**, *193*, 216–228. [CrossRef] [PubMed]

253. Grob, V.; Moline, P.; Pfeifer, E.; Novelo, A.R.; Rutishauser, R. Developmental morphology of branching flowers in *Nymphaea prolifera*. *J. Plant Res.* **2006**, *119*, 561–570. [CrossRef] [PubMed]

254. Arber, A. The Gramineae. In *A Study of Cereal, Bamboo, and Grass*; (Reprint 1965 by J. Cramer, Weinheim); Cambridge University Press: Cambridge, UK, 1934.

255. Stebbins, G.L.; Yagil, E. The morphogenetic effects of the hooded gene in barley. I. The course of development in hooded and awned genotypes. *Genetics* **1966**, *54*, 727–741.

256. Yagil, E.; Stebbins, G.L. The morphogenetic effects of the hooded gene in barley. II. Cytological and environmental factors affecting gene expression. *Genetics* **1969**, *62*, 307–319.

257. Müller, K.J.; Romano, N.; Gerstner, O.; Garcia-Maroto, F.; Pozzi, F.; Salamini, F.; Rohde, W. The barley *Hooded* mutation caused by a duplication in a homeobox gene intron. *Nature* **1995**, *374*, 727–730. [CrossRef]

258. Roig, C.; Pozzi, C.; Santi, L.; Müller, J.; Wang, Y.; Stile, M.R.; Rossini, L.; Stanca, M.; Salamini, F. Genetics of barley *Hooded* suppression. *Genetics* **2004**, *167*, 439–448. [CrossRef]

259. Williams-Carrier, R.E.; Lie, Y.S.; Hake, S.; Lemaux, P.G. Ectopic expression of the maize *knl* gene phenocopies the *Hooded* mutant of barley. *Development* **1997**, *124*, 3737–3745.

260. Bommert, P.; Whipple, C. Grass inflorescence architecture and meristem determinacy. *Semin. Cell Dev. Biol.* **2018**, *79*, 37–47. [CrossRef]

261. Reinhardt, D. Phyllotaxis—A new chapter in an old tale about beauty and magic numbers. *Curr. Opin. Plant Biol.* **2005**, *8*, 487–493. [CrossRef] [PubMed]

262. Loiseau, J.-E. *La Phyllotaxie*; Masson: Paris, France, 1969.

263. Zagorska-Marek, B. Phyllotactic patterns and transitions in *Abies balsamea*. *Can. J. Bot.* **1985**, *63*, 1844–1854. [CrossRef]

264. Rutishauser, R.; Peisl, P. Phyllotaxy. In *Encyclopedia of Life Sciences*; Macmillan Publishers Ltd.: Basingstoke, UK, 2001.

265. Rutishauser, R. Plastochrone ratio and leaf arc as parameters of a quantitative quantitative phyllotaxis analysis in vascular plants. In *Symmetry in Plants*; Jean, R.V., Barabé, D., Eds.; World Scientific: Singapore, 1998; pp. 171–212.

266. Rutishauser, R. *Acacia* (wattle) and *Cananga* (ylang-ylang): From spiral to whorled and irregular (chaotic) phyllotactic patterns—A pictorial report. *Acta Soc. Bot. Pol.* **2016**, *85*, 3531. [CrossRef]

267. Strauss, S.; Lempe, J.; Prusinkiewicz, P.; Tsiantis, M.; Smith, R.S. Phyllotaxis: Is the golden angle optimal for light capture? *New Phytol.* **2020**, *225*, 499–510. [CrossRef]

268. Barabé, D.; Lacroix, C. *Phyllotactic Patterns: A Multidisciplinary Approach*; World Scientific Publ.: Singapore, 2020.

269. Bartlett, M.E.; Thompson, B. Meristem identity and phyllotaxis in inflorescence development. *Front. Plant Sci.* **2014**, *5*, 508. [CrossRef]

270. Cutter, E.G. The inception and distribution of flowers in the Nymphaeaceae. *Proc. Linn. Soc. Bot.* **1961**, *172*, 93–100. [CrossRef]

271. El, E.S.; Remizowa, M.V.; Sokoloff, D.D. Developmental flower and rhizome morphology in *Nuphar* (Nymphaeales): An interplay of chaos and stability. *Front. Cell Dev. Biol.* **2020**, *8*, 303. [CrossRef]

272. Stevenson, D.W. Observations on phyllotaxis, stelar morphology, the shoot apex and gemmae of *Lycopodium lucidulum* Michaux (Lycopodiaceae). *Bot. J. Linn. Soc.* **1976**, *72*, 81–100. [CrossRef]

273. Rutishauser, R. Ever since Darwin: Why plants are important for evo-devo research. In *Perspectives on Evolutionary and Developmental Biology: Essays for Alessandro Minelli*; Fusco, G., Ed.; Padova University Press: Padova, Italy, 2019; pp. 41–55.

274. Grob, V.; Pfeifer, E.; Rutishauser, R. Sympodial construction of Fibonacci-type leaf rosettes in *Pinguicula moranensis* (Lentibulariaceae). *Ann. Bot.* **2007**, *100*, 857–863. [CrossRef]

275. Endress, P.K. Chaotic floral phyllotaxis and reduced perianth in *Achlys* (Berberidaceae). *Bot. Acta* **1989**, *102*, 159–163. [CrossRef]

276. Smith, R.; Guyomarc'h, S.; Mandel, T.; Reinhardt, D.; Kuhlemeier, C.; Prusinkiewicz, P. A plausible model of phyllotaxis. *Proc. Natl. Acad. Sci. USA* **2006**, *103*, 1301–1306. [CrossRef] [PubMed]

277. Yin, X. Phyllotaxis: From classical knowledge to molecular genetics. *J. Plant Res.*. under review.

278. Fierz, V. Aberrant phyllotactic patterns in cones of some conifers: A quantitative study. *Acta Soc. Bot. Pol.* **2014**, *84*, 261–265. [CrossRef]

279. Gola, E.M.; Jernstedt, J.A.; Zagórska-Marek, B. Vascular architecture in shoots of early divergent vascular plants. *New Phytol.* **2007**, *174*, 774–786. [CrossRef]

280. Yin, X.; Meicenheimer, R.D. The ontogeny, phyllotactic diversity, and discontinuous transitions of *Diphasiastrum digitatum* (Lycopodiaceae). *Am. J. Bot.* **2017**, *104*, 8–23. [CrossRef]

281. Hofmeister, W. *Allgemeine Morphologie der Gewächse*; W. Engelmann: Leipzig, Germany, 1868.

282. Imaichi, R.; Hiyama, Y.; Kato, M. Leaf development in the absence of a shoot apical meristem in *Zeylanidium subulatum* (Podostemaceae). *Ann. Bot.* **2005**, *96*, 51–58. [CrossRef]

283. Griffiths, P.E.; Stotz, K. Developmental systems theory as a process theory. In *Everything Flows: Towards a Processual Philosophy of Biology*; Nicholson, D.J., Dupré, J., Eds.; Oxford University Press: Oxford, UK, 2018; pp. 225–245. [CrossRef]

284. Jaeger, J.; Monk, N. Everything flows: A process perspective on life. *EMBO Rep.* **2015**, *36*, 1064–1067. [CrossRef]

285. Jaeger, J. Dynamic structures in evo-devo: From morphogenetic fields to evolving organisms. In *Perspectives on Evolutionary and Developmental Biology. Essays for Alessandro Minelli*; Fusco, G., Ed.; Padova University Press: Padova, Italy, 2019; pp. 335–355.

286. Scheres, B.; McKhann, H.I.; van den Berg, C. Roots redefined: Anatomical and genetic analysis of root development. *Plant Physiol.* **1996**, *111*, 959–964. [CrossRef]

287. Louarn, G.; Song, Y. Two decades of functional-structural plant modelling: Now addressing fundamental questions in systems biology and predictive ecology. *Ann. Bot.* **2020**, *126*, 501–509. [CrossRef]
288. Keller, E.F. *A Feeling for the Organism. The Life and Work of Barbara McClintock*; Macmillan: New York, NY, USA; San Francisco, CA, USA, 1983.
289. Arber, A. Goethe's Botany: The Metamorphosis of Plants (1790) and Tobler's Ode to Nature (1782). *Chron. Bot.* **1946**, *10*, 63–126.

Publisher's Note: MDPI stays neutral with regard to jurisdictional claims in published maps and institutional affiliations.

 *philosophies*

Article

How Do Living Systems Create Meaning?

Chris Fields [1,*] and Michael Levin [2]

[1] Independent Researcher, 11160 Caunes Minervois, France
[2] Allen Discovery Center at Tufts University, Medford, MA 02155, USA; michael.levin@tufts.edu
* Correspondence: fieldsres@gmail.com; Tel.: +33-6-44-20-68-69

Received: 26 September 2020; Accepted: 6 November 2020; Published: 11 November 2020

Abstract: Meaning has traditionally been regarded as a problem for philosophers and psychologists. Advances in cognitive science since the early 1960s, however, broadened discussions of meaning, or more technically, the semantics of perceptions, representations, and/or actions, into biology and computer science. Here, we review the notion of "meaning" as it applies to living systems, and argue that the question of how living systems create meaning unifies the biological and cognitive sciences across both organizational and temporal scales.

Keywords: active inference; attention; development; evolution; language; memory; pragmatics; reference frames; scale-free cognition; self; stigmergy

1. Introduction

The "problem of meaning" in biology can be traced to the Cartesian doctrine that animals are automata, and that humans are automata (bodies) controlled by spirits (minds). As Descartes required minds to generate meaning, meaning under this doctrine is both a strictly human and a strictly psychological phenomenon. The general turn from metaphysics toward language in mid-20th-century philosophy further reinforced this Cartesian division by localizing the study of meaning within the study of language, understood as human natural language characterized by a recursive grammar, and arbitrarily-large lexicon, and an associated collection of interpretative practices. Human cognition was widely assumed within the representationalist cognitive science that largely replaced behaviorism from the 1960s onward to replicate this tripartite structure of public natural language, either because it was implemented in an underlying "language of thought" or "mentalese" having a similarly expressive syntax, semantics, and pragmatics [1], or because it was implemented by natural language itself [2]. Cognitive abilities considered to be uniquely human, including symbolic "Process-2" cognition [3], analogical reasoning [4], and "mental time travel" to the past via episodic memory or the future via prospective memory [5], in particular, were all considered to be functionally dependent upon language.

During this same time, however, studies in comparative psychology demonstrated robust communication abilities, and related abilities including tool use and cultural transmission of knowledge, in animals that appear, on all assays thus far, to lack a recursive syntax [6]. Superficially similar communication abilities have since been demonstrated in microbes [7] and plants [8]. If "meaning" is restricted to the full combination of syntactic, semantic, and pragmatic aspects of meaning characteristic of human languages, the communications sent and received by these nonhuman organisms must be regarded as devoid of meaning, as Descartes presumably would have regarded them, and as communications sent and received by artificial intelligence (AI) systems are regarded by many today [9–12]. Hauser, Chomsky, and Fitch [6] suggest, on the contrary, that the faculty of language can be construed more broadly, and that both sensory–motor and conceptual–intentional aspects of "meaning" can be dissociated from the recursive syntax with which they are coupled in human languages. This dissociation allows nonhuman communication abilities to be viewed as meaningful even though not "grammatical" in the sense of having a compositional semantics enabled by a recursive syntax [13–15]. It is this broader, non-Cartesian conception of meaning with which we will be concerned in what follows.

Opposition to Cartesian assumptions and to human-like, grammatical language as paradigmatic in cognitive science has coalesced over the past three decades into the embodied-embedded-enactive-extended cognition (4E; see [16] for an overview) and biosemiotic (see [17] for an overview) movements. Both take considerable inspiration from Maturana and Varela's theories of autopoesis [18] and embodiment [19], and both are broadly consistent with cognition being a fundamental characteristic of living systems [14,15,20 24]. There is, however, considerable variation among non-Cartesian conceptions of meaning. Following Gibson [25], ecological realists locate meanings in the organism- or species-specific environment, where they have an effectively nomological function [26,27]. "Radical" embodied cognition locates meanings in the structure and dynamic capabilities of the body [28], while enactive cognition locates meanings in the organism's inclination and ability to act on the environment [29]; in both cases, meanings characterize structural and functional, but in most cases explicitly non-representational, capacities of an embodied system (see also [30,31] for reviews of these approaches from an AI/robotics perspective, see [32] for a dynamical-systems approach, and see [33] for a 4E approach that can be interpreted representationally). Biosemiotic approaches are, in contrast, concerned with physically implemented representations ("signs") that carrying meaning at one or more scales of biological organization.

Despite their differences, both Cartesian and non-Cartesian approaches to cognition generally localize meaning to individual organisms. Even culturally shared meanings, e.g., the particular syntactic markers, word meanings, and pragmatics shared by competant users of some particular human natural language, are assumed to be individually comprehended by every member of the language-using community. Such meanings are generally acknowledged to have a developmental history, typically involving learning in a community context, but outside of evolutionary psychology, they are seldom considered to have significant evolutionary histories. This focus on the individual and on learning is understandable from the perspective of (narrowly-construed) language and its pervasive influence on 20th-century thought. While the question of how the human language system, specifically its syntactic components [34], evolved is of theoretical interest, it is hard to imagine how individual, conventionalized word meanings, e.g., "<cat> means cat" could have significant evolutionary histories.

In contrast to this focus on individuals and learning, we advance in this paper a deeply evolutionary approach to meanings and suggest, consonant with the theme of this Special Issue, that the construction of meanings presents a common and pressing question for all disciplines in the Life Sciences. We focus, in particular, on three questions that are foundational to the study of meaning:

1. How do living systems distinguish between components of their environments, considering some to be "objects" worthy of attention and others to be "background" that is safely ignored?
2. How do living systems switch their attentional focus from one object to another?
3. How do living systems create and maintain memories of past events, including past perceptions and actions?

These questions all presuppose an answer to a fourth question:

4. How do living systems reference their perceptions, actions, and memories to themselves?

We use the term "living system" instead of "organism" in formulating these questions to emphasize their generality: we ask these questions of living systems at all scales, from signal transduction pathways within individual cells to communities, ecosystems, and extended evolutionary lineages. In addressing these questions, we build on our previous work on memory in biological systems [35] and on evolution and development as informational processes operating on multiple scales [36,37].

In what follows, we address these questions in turn, reviewing in each case both theoretical considerations and empirical results for systems at organizational scales ranging from molecular interaction networks to the evolutionary history of the biosphere as a whole, and time scales ranging from the picosecond (ps) scale of molecular conformational change to the billion-year (1000 MY) scale of planetary-scale evolutionary processes. To render these questions precise, we borrow from physics the idea of a *reference frame* (RF), a standard or coordinate system that assigns units of measurement to observations, and thereby renders them comparable with other observations. Clocks, meter sticks, and standardized masses are canonical physical RFs [38]. We show in Section 2 that organisms must implement internal RFs to enable comparability between observations, and make explicit the role of RFs in predictive-coding models of cognition [39–45]. The three sections that follow address questions (1)–(3) above, beginning in Section 3 with the function of RFs in object segregation, categorization, and identification. These mechanisms, which are best characterized in human neurocognitive networks, reveal how living systems determine what is potentially significant in their environments, and allow a basic characterization of the limitations on what living systems are capable of considering significant. We then consider in Section 4 the question of attention, a phenomenon also best characterized in humans. We suggest that the fundamental dichotomy between proactive and reactive attention systems found in mammals [46,47] can be extended to all scales, a suggestion consistent with Friston's characterization of proactive and reactive modes in active inference systems [41,42]. The question of memory is addressed in Section 5, where we examine the neurofunctional concept of an engram [48,49] and how it relates to biological memories at other scales [35]. Specifically, we ask whether the phenomenon of memory change during reconsolidation after reactivation [50,51] characterizes biological memories in general. We suggest that, while memories may be stored either internally or externally, the mechanisms and consequences of access are the same. The distinction between "individual" and "public" memories is, therefore, unsustainable. Object identification, attention, and memory are brought together in Section 6 in the construction of a self-representation to which meanings are associated. We explicitly consider the self-representations implemented by humans [52–55], and suggest that locating the self-representations operative in other organisms and in sub- or supra-organismal systems represents a key challenge to the Life Sciences [24]. We integrate these themes from an evolutionary perspective in Section 7, suggesting that meaning is itself a multi-scale phenomenon that characterizes all living systems, from molecular processes to Life on Earth as a whole. The fundamental goal of the Life Sciences is, from this perspective, to understand how living systems create meaning.

2. Meanings Require Reference Frames

Bateson famously defined a "unit" of information as a "difference which makes a difference" ([56] p. 460). As Roederer points out, information so defined is actionable or pragmatic; it "makes a difference" for what an organism can *do* [57]. It is, therefore, information that is meaningful to the organism in a context that requires or affords an action, consistent with sensory-motor meaning being the most fundamental component of language as broadly construed [6]. It is in this fundamental sense that meaning is "enactive" [19].

A "difference which makes a difference" must, clearly, be recognized as a difference, i.e., as being different from something else. The "something else" that allows differences to be recognized is an RF. Choosing an RF is choosing both a *kind of difference* to be recognized, e.g., a difference in size, shape, color, or motion, and a *specific reference value*, e.g., this big or that shape, that the difference is a difference from. Any discussion of differences assumes one or more RFs. Our goal in this section is make the nature of these RFs fully explicit. With an understanding of how specific kinds of RFs enable specific kinds of meanings, we can approach questions about the evolution, development, and differentiation of meanings as questions about the evolution, development, and differentiation of RFs.

2.1. System—Environment Interaction as Information Exchange

While definitions of "life" and "living system" are numerous, varied, and controversial [58–62], all agree that every living system exists in interaction with an environment. If Life on Earth as a whole is considered a living system [35,37,63,64], its environment is by definition abiotic; for all other living systems (on Earth), the environment has both abiotic and living components. Definitions of life agree, moreover, that at any given instant t, the state $|S\rangle$ of a living system S is distinct, and distinguishable, from the state $|E\rangle$ of its environment E (we borrow Dirac's $|\cdot\rangle$ notation for states from quantum theory: $|X\rangle$ is the state of some system X). This condition of state distinguishability, called "separability" in physics, guarantees that the interaction between a living system and its environment can be viewed, without loss of generality, as an exchange of classical information [65–67]. This exchange is symmetric at every instant t (technically, within every time interval Δt small enough that neither system nor environment significantly changes its size or composition during Δt): every bit (binary unit) of information that the living system obtains from its environment at t is balanced by a bit transferred by the living system to its environment at t. We are interested in interactions that provide S with actionable information about E; such interactions are thermodynamically irreversible and, as such, have a minimum energetic cost of $\ln 2\, k_B T$, k_B Boltzmann's constant and T temperature [68,69]. In this case, incoming bits can be considered "observations" and outgoing bits can be considered "actions", bearing in mind that in this classical, thermodynamic sense of information, obtaining free energy from the environment is "observing" it and radiating waste heat into the environment is "acting" on it. As the value of k_B in macroscopic units is small (1.38×10^{-23} Joules/Kelvin), bit-sequence exchange can appear continuous when temperatures are high and time is coarse-grained, as is typically the case in biological assays.

Under these conditions, we can ask how much a living system S can learn, i.e., what information S can obtain by observation, about the internal structure or dynamics of its environment E, and, conversely, how much E can learn about the internal structure or dynamics of S. The answer is that what can be learned at t is strictly limited to the classical information actually exchanged at t. This information is, for any pair of finite, separable physical systems, regardless of their size or complexity, strictly insufficient to fully determine the internal structure or dynamics of either interaction partner [70,71]; see [72,73] for informal discussions of this point. The information obtainable by either party by observation is, therefore, conditionally independent of the internal structure or dynamics of the other party; these can, in principle, vary arbitrarily without affecting the observations obtained at any given t. The interaction between any S and its E thus satisfies the Markov blanket (MB) condition: the information exchanged at t is the information encoded by an identifiable set of "boundary" states that separate the internal states of S from the internal states of E [33,41,42,74,75].

The idea that the observational outcomes obtainable by any observer are conditionally independent of the internal structure and dynamics of the observed environment has, of course, a long philosophical history, dating at least from Plato's allegory of the cave and forming the basis of Empiricist philosophy since Hume [76]. This idea challenges any objective ontology; pairwise interactions between separable systems—between any S and its E—are provably independent of further decompositions of either system in both quantum and classical physics [72]. As Pattee [77] puts it, the "cuts" that separate the observed world of any system into "objects" are purely epistemic and hence relative to the system making the observations. Understanding what "objects" S "sees" as components of its E thus requires examining the internal dynamics of S. These internal dynamics, together with the system—environment interaction, completely determine what environmental "objects" S is capable of segregating from the "background" of E and identifying as potentially meaningful. Whether it is *useful* to S to segregate "objects" from "background" in this way is determined not by the internal dynamics of S, but by those of E. Meaning is thus a game with two players, not just one. It is in this sense that it is fundamentally "embedded" (again see [19]). In the language of evolutionary theory, it is always E that selects the meanings, or the actions they enable that have utility in fact for S, and culls those that do not.

2.2. Meaning for Escherichia coli: Chemotaxis

What does a living system consider meaningful? Before turning to any division of the environment into objects and background, let us consider meanings assigned to the environment as a whole.

Bacterial chemotaxis has long served as a canonical example of approach/avoidance behavior and hence of the assignment of valence to environmental stimuli that "make a difference" to the bacterium (see [14,78] for recent reviews). Chemotaxis receptors respond to free ligand and have sufficient on-receptor short-term memory to determine local ligand gradients (see [79] for examples of cells that locally amplify gradients). The fraction of receptors bound at t indicates an environmental state, either $|good\rangle$ or $|bad\rangle$, at t and directly drives either "approach" or "avoid" motility. No division of E into "objects" is either possible or necessary in this system.

While it does not require object segregation, the assignment of valence to $|E\rangle$ during bacterial chemotaxis does require an RF to distinguish between $|good\rangle$ and $|bad\rangle$. In *E. coli*, this RF is defined by the default phosphorylation state R_0 of the messenger protein CheY, i.e., by the default value of the concentration ratio $R = [\text{CheY-P}]/[\text{CheY}]$ [78]. Values of R greater or less than R_0 encode the "pointer states" (again borrowing terminology from physics) $|R\rangle = |good\rangle$ or $|R\rangle = |bad\rangle$ and induce approach or avoidance, respectively. The value R_0 is "set" by the balance between kinases and phosphatases acting on CheY (with CheY acetylation as a secondary modulator; see [80] for discussion). This balance is, in turn, dependent on the overall state of the *E. coli* gene regulation network (GRN) as discussed further in Section 6 below.

As noted earlier, the concept of an RF was originally developed within physics to formalize the pragmatic use of clocks, measuring sticks, and other tools employed to make measurements [38]. In *E. coli*, however, we see an *internal state* being employed as an RF. This situation is completely general: any use of an external system as an RF presupposes the existence of an internal state that functions as an RF [81]. An external clock, for example, can only register the passage of time for an observer with an internal RF that enables comparing the currently-observed state of the clock to a remembered past state. The sections that follow will illustrate this for progressively more complex RFs. A formal construction of internal RFs given a quantum-theoretic description of the system-environment interaction is provided in [66].

2.3. Implementing RFs Requires Energy

The CheY phosophorylation system in *E. coli* effectively encodes one bit of information: the pointer state $|R\rangle$ can be either $|good\rangle$ or $|bad\rangle$. Encoding this bit requires energy: at least $\ln 2 \, k_B T$ as discussed above. The energy needed is in fact much larger than this, as the protein CheY itself, the kinases and phosphatases that act on it, and the receptors that provide input to the CheY phosphorylation/dephosphorylation cycle must all be synthesized and eventually degraded, and the intracellular environment that supports these interactions must be maintained. All of these processes are thermodynamically irreversible. The energy to drive these processes must be supplied by *E. coli*'s metabolic system. It derives, ultimately, from the external environment. The external environment also absorbs, again irreversibly, the waste heat that these biochemical processes generate.

The energetics of the CheY system illustrate a general point: implementing internal RFs requires energetic input from the environment. This energetic input is necessarily larger than the energy required to change the pointer state associated with the RF [66]. Any RF is, therefore, a dissipative system that consumes environmental free energy and exhausts waste heat back to the environment. Every RF an organism implements requires dedicated metabolic resources. In mammals, for example, the total energy consumption of the brain scales roughly linearly with the number of neurons, indicating an approximately constant per-neuron energy budget [82]. Energy usage is highest in cortical neurons, roughly $10^{11} \, k_B T$ per neuron per second. Using rough estimates of 10,000 synapses per neuron receiving data at 10 Hz, on the order of $10^6 \, k_B T$, spread out over multiple maintenance, support, and internal computational processes, is required to process one synaptically-transferred bit.

2.4. RFs Are State-Space Attractors

Following a perturbation, the *E. coli* phosphorylation ratio R returns to its default value R_0. Stability against perturbations within some suitable dynamic range is clearly required for any state to serve as an RF. Hence, RFs are state-space attractors. Its dimensionality as an attractor defines the dimensionality of the "difference" the RF makes in the behavior of the system's state vector. An RF is, on this definition, a point or collection of points (e.g., a limit cycle) in the state space that separates state-vector components associated with different behaviors. In the case of $|R\rangle$, the dimensionality is one, corresponding to 1 bit of information: the value of R, either greater or less than R_0, determines whether the flagellum spins clockwise (approach) or counter-clockwise (avoid).

2.5. RFs Set Bayesian Expectations

As Friston and colleagues have shown [41,42,74,75], the behavior of even unicellular organisms can generically be described in Bayesian terms. In this language, an RF such as the default value R_0 of [CheY-P]/[CheY] defines an expectation value, i.e., a "prior probability" for the sensed state of a valence-neutral environment. Avoidant chemotaxis in *E. coli* is an instance of active inference that restores a default expectation; approach chemotaxis is expectation-dependent goal pursuit. Short-term memory for environmental state is implemented by receptor methylation [78], while longer-term revision of expectations is implemented by modulating R_0; we discuss the roles of RFs as memories in active inference systems more generally in Section 5 below.

2.6. Only Meaningful Differences Are Detectable

When interaction is viewed as information exchange and meaningful differences are viewed as physically encoded by RFs, two aspects of meaning in biological systems become clear:

1. Detecting differences with respect to internal RFs is energetically expensive.
2. A difference is detected when at least one bit flips—at least one component of a pointer state changes.

Any *detectable* difference, therefore, *makes a difference*, both to the value of an "informative" pointer state and to the energy budget of the organism doing the detecting. Employing Bateson's criterion, all detectable differences are meaningful. Organisms do not waste energy acquiring information that is not actionable. This is not a choice, or a matter of optimization: acquiring information without changing state, and thereby acting on the environment, is physically impossible.

2.7. The Evolution of Meaning Is the Evolution of RFs

Consistent with its primarily biochemical, as opposed to mechanical or manipulative, mode of operating within its world [83], *E. coli* encodes RFs and associated pointer states, such as R_0 and $|R\rangle$ that can be interpreted as *tastes*. While *E. coli* is capable of locomotion using its flagellum, chemotaxis is opportunistic. The tumbling "avoid" motion due to counterclockwise flagellar rotation is directionally random; *E. coli* does not appear to encode any stable spatial RF. Pilus extension during "mate seeking" in *E. coli* similarly appears directionally random, and is limited in distance by the pilus polymerization mechanism [84]. The distal tip of the pilus carries an adhesin, but it is unknown whether it is specifically recognized by the recipient cell; hence it is unknown whether *E. coli* encodes an RF for "mate" or "conspecific". The *E. coli* cell cycle is opportunistic, and it has no circadian clock (although it is metabolically capable of supporting a heterologous clock [85]); hence, *E. coli* appears to have no endogenous time RF. The world of "taste" may, for *E. coli*, be the whole world.

How did evolution get from systems like *E. coli* that are restricted to tasting an environment that they may move in, but do not otherwise detect or represent, to systems like humans equipped with RFs for three-dimensional space and linear time, as well as RFs capable of identifying thousands of individual objects and tracking their state changes along multiple dimensions? This is the fundamental question for an evolutionary theory of cognition, one that goes far beyond what is standardly called "evolutionary psychology" [86,87]; see [88] for one way of asking this broader question. Addressing this question, we propose, requires understanding the evolution of RFs. As with other evolutionary questions, the available empirical resources are the implementations of RFs across a phylogenetically broad sample of extant organisms and the developmental and/or regenerative processes that construct these RFs.

Consistent with their role in detecting and interpreting states of the external environment, and as discussed in Section 6 below, the internal environment as well, one can expect cellular-scale RFs to be implemented by signal transduction pathways. One approach to the question of RF evolution is, therefore, to understand the evolution of signal transduction pathways (e.g., two-component pathways in microbes [89], or the Wnt [90] or MAPK [91] pathways in eukaryotes; see [92] for a general review). Understanding signal transduction in a "bag of molecules" is, however, not enough: the evolution of the organizing framework provided by the cytoplasm, cytoskeleton, and membrane—what we have previously termed the "architectome" [35]—also contributes to the evolution of RFs, particularly spatial RFs as discussed in Section 3 below. As systems become multicellular, we can expect RFs to be implemented by cell–cell communication systems including paracrine [93,94], endocrine [95,96], non-neural bioelectric [97–99], and neural [100–102] systems. Hence, the question of RF evolution incorporates the questions of morphological and physiological evolution, and, at larger scales, the evolution of symbiotic networks (including holobionts [103,104]), communities, and ecosystems [36,37]. At every level, RFs specify actionability and therefore meaning.

3. How Are Objects Segregated and Identified?

3.1. From Multi-Component States to Objects

While the chemotactic pathway in *E. coli* shows no evidence of specific object recognition as opposed to environmental state recognition, microbial predators such as *Myxococcus xanthus* appear capable of recognizing and differentially responding to both kin and non-kin conspecifics as well as a wide range of bacterial and fungal prey species (see [7,105] for recent reviews). While the RFs that distinguish cells to be killed (non-kin conspecifics or prey) from cells to be cooperated with (kin conspecifics) have not been fully characterized, the general structure of microbial sensor and response systems [14] suggest that default phosphorylation ratios analogous to R_0 are likely candidates.

Does *M. xanthus* represent conspecifics or prey as separate objects, or are these just different, possibly localized, "features" of its environment? When a traveling paramecium encounters a hard barrier, backs up, and goes around it, does it represent the barrier as an object, or just a "hard part" of the environment? When an amoeba engulfs a bacterium, is the bacterium an object, or just "some of" the environment that happens to be tasty? These questions cannot currently be answered, and the first organism (more properly, lineage) capable of telling an "object" from a feature of its environment remains unknown. From an intuitive perspective, an ability to distinguish "objects" from the environment in which they are embedded may come with the ability to specifically manipulate them. The sophisticated, stigmergic wayfinding signals used by social insects may be processed as environmental features, but eggs, prey items, agricultural forage, building materials, enemies, and even the dead bodies of colony-mates are manipulated in specific, stereotypical ways [106,107] and may be represented as objects. This association between manipulability and objecthood appears to hold for human infants [108,109]; that even artificial constructs such as road networks are intuitively considered "features" suggests that it holds for adult humans as well. Concluding that this association holds only for relatively advanced multicellular organisms may, however, not be justified; cells migrating within a multicellular body leave stigmergic messages for later-migrating cells to read [110,111] and cells of many types inspect their neighbors individually, killing those that fail to meet fitness criteria [112,113].

3.2. Objects as Reference Frames

Unlike *M. xanthus*, which has separate, functionally dissociable detection systems and hence separate RFs for "prey" and "edible prey components", cephalopods, birds, and mammals are capable of both categorizing objects and identifying objects as persistent individuals across changes in their states. This ability allows distinct meanings to be assigned to the same object at the category, individual, and state levels of specificity. The state-independent components of an object appear, in this case, to serve collectively as an RF for a categorized individual during a perceptual encounter; the state-independent components of a persistent object similarly appear to serve as an RF for individual re-identification. If this is the case, the state-independent meaning of a categorized individual can be identified with the set of actions, including inferences, that it affords as a recognizable object distinct from the "rest of" the environment.

Object recognition has been studied most intensively in mammals, particularly humans, and is best understood in the visual modality (see [114–118] for general reviews, see [119] for a comparison of haptic and visual modalities, and see [120] for pointers to work in other mammals and birds). Object category recognition ("categorization"), e.g., recognizing a table as a table, as opposed to a specific, individual table encountered previously, is primarily feature-based, with "entry-level" categories corresponding to early-learned common nouns, e.g., "table" or "person" learned first and processed fastest [117]. Identifying specific individual objects, e.g., specific individual people, is also feature based, but requires the use of context and causal-history information in addition to features if similarly-featured competitors or opportunities for significant feature change are present [121,122]. The object recognition process generates an "object token" [115], an excitation pattern in medial temporal cortex that is maintained over the course of a perceptual encounter (i.e., 10 s to 100 s of seconds) and that can be, but is not necessarily, encoded more persistently as a component (an engram [48,49] or invariant [123]) of an episodic memory. Identifying a categorized object as a specific, known individual requires linking its current object token to at least one episodic-memory encoded object token.

An object that never changes state and hence affords no actions is not worth recognizing (highly symmetric objects are an exception, e.g., [124]). An object that does change state—even if it just changes its position in the body-centered coordinates (see below) of a mobile observer—must have features that do not change to enable recognition. The collection of unchanging features serve as a reference against which variation in the changing feature—the pointer state of interest—can be tracked. They therefore constitute an RF for the object, at least during that perceptual encounter [66,81]. Such RFs may be extraordinarily rudimentary, and hence exploitable by competitors, predators, or potential prey through the use of mimicry, camouflage, lures, or other tactics. How humans or other animals construct stable object tokens out of sequences of encounter-specific RFs is not known, and constitutes a major open problem in cognitive and developmental psychology and in developmental and applied robotics [118]. Identifying individual objects as such across significant gaps in observation during which previously-stable features may have changed requires solving the Frame problem [125]. This problem is formally undecidable [126]; hence, only heuristic solutions are available. What heuristics are actually used by humans and how they are implemented remain unknown.

Consider again the use of an object such as a clock as an external RF. Not only does this require an internal time RF, it requires an RF to identify the clock as such, and indeed as the *same* clock that was consulted earlier. The clock can only be employed as a shared external RF by agents that share these internals RFs as well [81]. Apart from speculations about consciousness (e.g., [127–129]), physicists have historically been reluctant to comment on the internal structures of observers [130]. By clarifying the role of RFs in identifying objects and making measurements, biology and cognitive science may make significant contributions to physics.

3.3. Embedding Objects in Space

Representing either environmental features or objects as having distinct locations requires an RF for space. All cells, and all multicellular organisms, come equipped with a two-dimensional surface on which spatial layout can be defined: the cell membrane(s) that face the external environment. Locations on this surface are stabilized by the cytoskeleton at the cellular scale, and by cell–cell interactions and macroscopic architectural components such as skeletons at the multicellular scale. Physical, biochemical, and/or bioelectric asymmetries impose body axes on these surfaces [35,99,131]. Even *E. coli* has a body axis, with chemoreceptors on one end and flagella on the other.

Asymmetric placement of receptors and effectors on the surface facing the environment effectively embeds them in a two-dimensional body-centered coordinate system. Localized processing of information binds it to this coordinate system, allowing downstream systems to know "where" a signal is coming from [132]. The extent to which cytosolic signal transduction is spatially organized, e.g., by the cytoskeleton, at the cellular level remains largely unknown except in the case of neurons, where specific functions including exclusive-or [133] have been mapped to specific parts of the cell. In the neural metazoa (Ctenophores, Cnidarians and Bilaterians), the spatial organization of signal transduction is largely taken over by neurons. The ability of neurons to provide high-resolution point-to-point communication allows the internal replication of body-surface coordinate systems, e.g., in the somatosensory homunculus [134] or retinotopic visual maps [135]. Such replication is essential to fine motor control.

Locomotion and manipulation by appendages appear necessary for adding a third, radial dimension to body-centered coordinates. Stigmergic coordinate systems, for example, are meaningless without locomotor ability. Building a representation of space requires exploring space, even peripersonally. The construction of the mammalian hippocampal space representation from records of paths actually traversed reflects this requirement [136]. How such path-based spatial representations are integrated with stigmergic, celestial, or magnetic-compass based coordinate systems, and whether such integration is required to enable, e.g., long-distance migration, remain largely unexplored.

"Placing" objects in space ties them to the body-centered spatial RF, and enables them to function as external RFs for the body. This ability appears evident in arthropods, cephalopods, and vertebrates, and may be more widespread, e.g., in plants [8]. Humans are capable of radically revising such RFs as new entities are characterized as objects; for example, the RF defined by cosmological space has been revised multiple times from the European Medieval era to the present [137].

3.4. Embedding Objects in Time

At the cellular level, the primary time RFs are cyclic: the cell cycle and the circadian clock [138–140]. Embedding events or objects in longer spans of time requires a linear time RF, i.e., an ordered event memory with sufficient capacity. Evidence of cellular memory [141] or cellular-scale learning (see Section 5) across cell-cycle or diurnal boundaries is insufficient, on its own, to the establish the ordering capability required for an internal linear time RF. Cells can use ordered external events to control sequential behavior over longer times, as in migration and fruiting-body development in slime molds, but this is also insufficient to indicate an internal linear time RF.

By "internalizing" events that are external to the individual cells involved, multicellular systems can implement linear time RFs, e.g., the linear time RF implemented by a tissue- or organism-scale developmental process. Interestingly, highly-structured anatomies and morphologies, and hence highly-structured developmental processes correlate, in animals, with the presence of nervous systems, i.e., they occur in the Ctenophores, Cnidarians, and Bilaterals [102]. While local cell density, cell-type distribution, or other micro-environmental features may serve as external temporal markers, at least some developmental processes appear to be intrinsically timed. Tail regeneration in *Xenopus*, for example, synchronizes with the developmental stage at tail amputation to produce a tail appropriate to the overall size of organism once regeneration is complete [142].

Cephalopod, bird, and mammalian nervous systems implement short-term linear time RFs that enable highly-structured, feedback-responsive, ordered behaviors, e.g., pursuit, defense, courtship, or nest-building that span minutes to days. Many longer-term behavior patterns, however, are tied to monthly or annual cycles, not to linear time. Hence, it is not clear how long a purely-internal, linear time RF can be sustained. Even in humans, long linear time is supported by external RFs, including communal memory enabled by language and stable environmental records such as calendars (see Section 5 below).

4. How Is Attention Switched between Objects?

4.1. Active Inference Requires Attention

Active inference, at its most basic, is a trade-off between acting on the environment to meet expectations and learning from the environment to modify expectations [41,42]. Even with only a single stimulus, and hence a single component for which environmental variational free energy (VFE) is measured, the importance—encoded in the theory as Bayesian precision—placed on input versus expected values affects the balance between expectation-meeting and expectation-updating. Action on the environment can, moreover, include both exploitative actions that meet current expectations and exploratory actions that increase the probability of learning [143]. Enacting these distinctions requires a prioritization or attention system. Spreading VFE across multiple distinguishable stimuli increases the need for attention to allocate exploitative, exploratory, or passive learning resources. Even in *E. coli*, competition among chemoreceptors for control of flagellar motion implements a rudimentary form of attention.

Attention reorients action, at least temporarily, from one stimulus or goal to another. Even in organisms with sufficient cognitive resources to pursue multiple goals or action plans simultaneously, including humans, the focus of attention is generally unitary (animals such as cephalopods with highly-distributed central nervous systems may be an exception; see [144]). One can, therefore, expect attentional control to localize on processors with high fan-in from sensors and high fan-out to effectors. At the cellular level, this is the defining characteristic of "bow-tie" networks [145]; the CheY system of *E. coli* provides a rudimentary example, and the integrative role of Ca^{2+} in eukaryotic cells [146] a more highly ramified one (see [147] for additional examples). A bow-tie network is, effectively, a means of "broadcasting" a control signal to multiple recipients. In the mammalian central nervous system, this is the function of the proposed "global neuronal workspace" (GNW) [148–152], an integrative network of long-range connections between frontal and parietal cortices and the midbrain. Significantly, the GNW is the proposed locus of conscious attentional control, whether proactive or reactive.

4.2. What Is the RF for Salience?

Being a target for attention requires salience. How does the significance, current or potential, of a stimulus for an organism become a marker for salience? If everything detectable is meaningful, what makes some detectable states or objects more meaningful than others?

E. coli provides one of the simplest examples of salience regulation: the induction of the *lac* operon [153]. Glucose is salient to *E. coli* by default; when glucose is absent, *lac* operon induction renders lactose salient. This simple example generalizes across phylogeny in three distinct but related ways:

1. Some states or objects are salient by default, e.g., threats, food sources, or mating opportunities.
2. Salience is inducible. Antigens induce antibody production, amplifying their salience. Removal of a sensory capability, e.g., sight, enhances the salience of phenomena detected by surviving senses, e.g., audition or touch.
3. Control of salience is distributed over multicomponent regulatory networks, e.g., GRNs or functional neural networks.

Control systems that regulate salience are, effectively, RFs for salience. They enhance or repress meaningfulness in a context-dependent way. What, however, defines a context?

While the notion of context is often treated informally, Dzhafarov and colleagues [154–156] have developed a formal theory of context-dependent observation, couched in the language of classical probability theory, that is general enough to capture the irreversibility (technically, non-commutativity) of context switches characteristic of quantum theory [157] (see [66,158] for a Bayesian implementation).

Key to this approach is the idea that contexts are defined by sets of observables that are, in typical cases, *not salient*. As an example, consider the stalk supporting the lure of an anglerfish [159]. The stalk (a modified fin spine) is visible and is a critical marker of context, but in a fatal case of inattentional blindness, is ignored by prey that attack the lure.

When context is defined in this salience-dependent way, the control of salience becomes relational, and therefore global. Salience-control networks can thus be expected to display the kind of global semantic dependencies exhibited by semantic-net representations of word or concept meaning [160] or indeed, fully-connected networks of any kind, e.g., hyperlink dependencies on the internet. Not surprisingly, the "salience network" of the human connectome is a high-level loop connecting insular and cigulate cortices with reward and emotion-processing areas in the midbrain, and it is strongly coupled to the GNW and to proactive and reactive attention systems [161].

4.3. Salience Allocation Differences Self-Amplify

Differences in salience allocation between lineages, or between individuals within a lineage, will clearly contribute to differences in resource exploitation. More interesting, however, is their contribution to differences in resource exploration. Mutants of *E. coli* deficient in glucose uptake or metabolism, for example, must find and exploit environmental niches that provide other metabolizable sugars. In general, differences in salience allocation will result in different experiential and hence learning histories. Differences in learning can be expected to generate, in turn, additional differences in salience allocation. These differences may lead, as in the case of *E. coli* lacking glucose transporters, to niche segregation and possibly eventual speciation.

Lateral gene transfer (LGT) provides a means of sharing DNA between distantly-related lineages and is ubiquitous in the microbial world [83]. Genes or operons providing novel metabolic abilities, e.g., antibiotic resistance, are often shared by LGT. Viewed from the perspective of salience allocation, LGT is a communication mechanism that enables a donor to alter, possibly radically, the allocation of salience and hence attention by the recipient. Compared to LGT, quorum sensing provides for faster communication and hence more responsive salience reallocation, typically only among near kin as in *M. xanthus* biofilms [7]. These mechanisms provide models for the ubiquitous use of inter-individual and even interspecies communication to induce changes in salience allocation in both plants and animals. Diversification of communication signals, and hence of means to negotiate salience, may be a substantial driver, not just a consequence, of speciation [162].

5. How Are Memories Stored and Accessed?

Recognizing, responding to, and communicating meanings all require memory. As discussed in [35], the spatial scales of biologically-encoded memories span at least ten orders of magnitude, and their temporal scales almost twenty orders of magnitude. The evolution of life can be viewed as the evolution of memory, not only at the genome scale, but at all scales.

5.1. Heritable Memories Encode Morphology and Function

Jacob and Monod end their famous paper on the *lac* operon with the following words ([153] p. 354):

The discovery of regulator and operator genes, and of repressive regulation of the activity of structural genes, reveals that the genome contains not only a series of blue-prints, but a coordinated program of protein synthesis and the means of controlling its execution.

The biological memory implemented by the genome, Jacob and Monod discovered, encodes structure *and* function. Evolutionary change, even when restricted to the level of the genome, can affect not only components, but how, when, where, and in response to what they are made. The increased efficiency with which evolution can explore morphological and functional space by copying and modifying genetic regulatory systems is the key insight of evo-devo [163,164]. It can be generalized to evolution at all scales [35–37].

Heritable memories require reproductive cells. In multicellular organisms in which most cells are reproductively repressed, developmental biology and regenerative, cancer, and general stem-cell biology provide, respectively, opportunities to study heritable memory as encoded by germline and non-germline reproductive cells. These memories are encoded on multiple substrates, from molecular structures and concentration ratios through architectome organization to cellular-scale bioelectric fields [35]. They are enacted by cell proliferation, differentiation, migration, cooperation, and competition, i.e., by building a morphology that functions in the specified way.

The available RFs of organisms such as *E.coli* are specified by heritable memories, e.g., the genes coding for pathway components such as CheY and its kinases and phosphatases and the cytoplasmic conditions that allow such components to function. Heritable memories specify many RFs in multicellular organisms, including humans; examples include photoreceptors responsive to specific frequencies and olfactory receptors responsive to specific compounds. Such memories may be genome independent, e.g., the body-axis polarity RFs specified by the sperm-entry point in ascidians [165] or by bioelectric polarity [166,167] or axon orientation [168] in regenerating planaria.

5.2. Experiential Memories and Learning

Learning is ubiquitous in animals. It has also been demonstrated in paramecia [169] and in slime molds such as physarum [170], and possibly in plants [171–173]; hence learning does not require neurons, let alone networks of neurons [15]. Learning can adjust the salience of stimuli, modify the responses to stimuli, and at least in birds and mammals, introduce novel categories and hence induce the construction of novel RFs. While in many cases learning requires multiple exposures or extended experience with a stimulus, birds and mammals at least are also capable of one-shot learning [174]. Episodic memories in humans, for example, are all results of one-shot learning: each records a single, specific event.

Experimental work in numerous invertebrate and vertebrate systems has contributed significantly to understanding the neural implementation of particular experiential memories. Such "engrams" may be implemented by single cells, local ensembles of cells, or extended networks connecting cell ensembles in different parts of the brain, and hence may be encoded by patterned activity at the intracellular up to the functional-network scale [48,49]. The dendritic trees of individual neurons, particularly cortical pyramidal cells, long-distance von Economo neurons, or cerebellar Perkinje cells, are complex temporal signal-processing systems [175,176] that may be individually capable of computations as complex as time-windowed exclusive-or [133]. The functionality of these signal processors is reversibly modulated bioelectrically over short times and epigenetically over longer times [177,178]. Hence, neurons have far greater functionality than the simple sum-threshold units employed in typical artificial neural networks (ANNs), as has been understood by designers of "neuromorphic" computing systems since the late 1980s [179]. Both individual neurons and extended networks of neurons can, therefore, be expected to implement "learning algorithms" far more complex than those designed for typical ANNs. Intriguing recent results suggest, for example, that some neurons are either intrinsically, or are primed by some (possibly epigenetic) mechanism to be, more excitable than neighboring cells, and that these cells are preferentially recruited into networks that encode memories [49]. Such a prior bias toward particular cells is reminiscent of genetic pre-adaptation [180,181], suggesting a deep functional analogy between genetically and neuronally encoded memories.

Stigmergic implementation of event memories is also ubiquitous, particularly in social insects, cephalopods, birds, and mammals. Here, memory encoding requires acting on the environment, either biochemically via pheromones or other scent markers, or mechanically via techniques ranging from nest or den construction to human architecture or writing. Learning, in these cases, includes learning to both write and read externally-encoded memories. These activities, clearly, require the internal implementation of RFs to assign semantics to both the actions involved in writing and the sensory inputs involved in reading. Comparative analyses of communication systems (e.g., [182–184]) and tool use (e.g., [185–188]) provide particularly promising avenues to explore both commonalities (e.g., the ubiquitous role of FoxP in communication systems [189]) and differences in the learning and encoding of these RFs.

5.3. Reporting, Reconsolidation, and Error Correction

Any behavior can be considered a "report" of one or more memories. Understanding which memories are being reported is clearly an experimental and interpretative challenge, even in the case of human verbal reports. More interesting in the present context, however, is the question of whether, and if so how, reporting a memory by somehow enacting its content alters that content. Memory content change on reconsolidation is well documented in the case of human episodic memories [50,51]. Can this be expected for memories across the board? If memories are generically unstable, what kinds of redundancy or error correction systems can be expected?

It is clear that reconsolidation generically effects revision in at least one other form of memory: object tokens. As objects change state, their identifying features may change; state changes accompanying growth in individual humans are an obvious example. Object tokens must, therefore, be continually updated [122]; while previous versions may be maintained in a "historical model" of the individual, they are recalled by chaining backwards from the current version. Within a nervous system, these are all dynamic processes that involve exciting some cells and inhibiting others, i.e., they involve the same cellular-scale changes that record memories in the first place. Reactivation (recall) and reconsolidation can, therefore, be expected to cause memory revision generically.

Humans employ both communication with others and external objects, including written documents and recorded images, to supplement individual memories and hence to provide both error correction and the possibility for restoring memories that have been degraded or lost. The maintenance of a shared language, for example, requires communication and benefits from recording and writing. Words (nouns) are markers for object tokens, and hence for object-identifying RFs. We can expect the maintenance of shared RFs for external objects to generically require both communication and shared external exemplars, e.g., "canonical" objects that all within a community agree are tables, tools, or most critically, in-group or out-group members. Members of a linguistic community must, moreover, be capable of recognizing that other members agree about the status of such objects and the words used to refer to them [190].

Redundancy and "community" based error correction are, effectively, methods for distributing memory across multiple, at least partially interchangable representations. We suggest that such distributed-memory methods are employed in analogous ways at all scales of biological organization, from double-stranded DNA and the maintenance of multiple copies of the genome in most eukaryotic cells, to quorum sensing and other means by which cells "negotiate" and "agree" by exchanging molecular or bioelectric signals, to the redundancy and handshaking mechanisms employed by neural, hormonal, and immune systems. Organisms capable of preserving experiential memories across events in which the brain is substantially remodeled (e.g., in insect metamophosis) or even completely regenerated following injury (e.g., in planaria) provide compelling examples of distributed memory [191]. While we do not, in general, know the "words" being used in these memories or the RFs with which they are associated, the languages of communication and information flow are so intuitively natural that they have been employed to describe such systems even in the absence of detailed computational analysis or specific functional analogies. We suggest that viewing these systems explicitly in terms of computational models, e.g., in terms of VFE minimization and Bayesian satisficing [41,42,74,75], will prove increasingly useful.

5.4. Reconsidering the Cognitive Role of Grammatical Language

Converging evidence from functional neuroscience and cognitive, evolutionary, and social psychology has led over the past decade to a substantial rethinking of the role of grammatical language, and indeed of the symbolic constructs of "good old-fashioned" AI (GOFAI) in general, in the implementation of human cognition. One strand of this rethinking focuses on the role uniquely played by language, and by the deliberative, inner-speech driven Process-2 cognition underlying public language use, in human communication. As Adolphs [190] emphasizes, the primary selection pressures in later human evolution were social. In such a setting, language serves not only as a means of cooperative communication, but of justification, pursuasion, and deception, including self-deceptive rationalization [192–195]. These uses of language suggest a role for language on the "surface" of cognition, above the level of, e.g., situational awareness or planning. The general, cross-domain association of expertise (including expertise in the use of language) with automaticity [196–198] and the difficulty of achieving a fully self-consistent definition of Process-2 cognition [199] similarly suggest a surface role for both Process-2 cognition and language. Chater [200] formulates this "flat" conception of cognition as "mental [i.e., cognitive and emotional] processes are *always* unconscious—consciousness reports answers, but not their origins ... we are only ever conscious of the results of the brain's interpretations—not the 'raw' information it makes sense of, or the intervening inferences" (p. 180, emphasis in original), a characterization that comports well with the MB condition [33], the idea that perception and action act, in all modalities, through an "interface" separating the cognitive system from the world [201], and the GNW architecture [148]. Indeed Process-2 cognition can be fully implemented by automated, Process-1 cognition in a GNW setting [202].

If language use is a modal, surface phenomenon implemented by domain-general, scale-free mechanisms such as hierarchical Bayesian satisficing [41,42], the need to identify specific neural implementations of syntactic or lexical symbols at the computational level [1,2,203] to explain language use diminishes, and possibly disappears altogether. Historically, the greatest barrier to domain-nonspecific models of language use has been recursion [6,204]. Recursive sequence learning and production is not, however, limited to language; visual processing [205] and motor planning [206,207] are now known to be both recursive and independent of language. These earlier-evolving systems may have been co-opted for language processing, a co-option supported by comparative phenotypic analysis of FoxP mutations [208]. The motor planning system is a canonical predictive-processing system [209] that can be described in terms of hierarchical Bayesian satisficing in a way largely analogous to visual processing [210]. Significantly, neuroimaging studies consistently localize numerical and mathematical reasoning, including abstract symbolic reasoning, to fronto-parietal networks that overlap visual motion-detection and motor areas, not language-processing (e.g., Broca's) areas [211,212]. As with grammatical sentence construction, whether abstract mathematical manipulations are implemented by a hierarchical Bayesian mechanism remains unknown.

6. How Do Living Systems Represent Themselves?

One of us (ML) has previously suggested that the "self" is usefully thought of in terms of a "cognitive light cone" indicating the horizon of the goal states the agent is capable of pursuing. The spatial extent demarcates the distance across which it is able to take measurements and exert effects, and the forward and backward temporal extents indicate the system's abilities to anticipate the future and recall the past, respectively [24]. The "past" here is the past of experienced events, and hence of memories resulting from learning. While many non-human animals clearly remember places, events, and social roles and some are capable of planning cooperative hunts or raids against neighboring populations, evidence for "mental time travel" to the past via episodic memory or the future via prospective memory is contested, and these are often regarded as human-specific [5]. It is not clear, in particular, whether any non-humans have a sufficiently flexible self-representation to represent their own past or future actions.

The status of the human self-representation also remains unclear. Electro-encephalographic (EEG) studies of experienced meditators localize the experienced emotional-agentive, narrative-agentive, and passive "witness" aspects of the self to right- and left-posterior and mid-frontal excitations, respectively [213]. The insular-cingulate-limbic loop involved in salience allocation is also involved in linking the interoceptive sense of a bodily self to both cognitive control and memory [52–55]. How this pragmatic self-representation relates to the "psychological" self characterized by beliefs, desires, ethical attitudes, and personality, and even whether this latter self exists, remain open questions [200,214,215].

It seems natural to view the pragmatic self representation as an RF, but what exactly does it measure, how is it implemented, and most critically, how do implementations at different scales relate? Levin [24,216] considers a self to be a stable unit of cooperative action, whether at the cellular, tissue, organ, organism, community, or even higher scale. At what scale does the self-representation become a distinct functional component, as it appears to be in humans, of the self it represents? Heat-shock and other stress-response proteins provide measurable indicators of cellular state to other cellular processes; one can, for example, consider the *E. coli* heat-shock system an RF for environmental stress. It is not, however, clear that any component of the metabolic and regulatory network of *E. coli* represents the state of the rest of the network in the way that the human insular-cingulate-limbic loop represents the state of the body to frontal control systems, or in the way that the somatosensory homunculus represents the body to the tactile-processing system. The functional architecture of *E. coli* involves many communicating components, but does not appear to involve an overall metaprocessor. Is metaprocessing, and hence the need for one or more separate self-RFs, part of the mammalian innovation of developing a cerebral cortex?

Cancer is widely considered a "rebellion" on the part of some cells against the cooperative requirements of the whole organism. Does the "new self" of a tumor emerge de novo, or is it latent, and somehow actively repressed, in well-behaved somatic cells? Does the immune system have specific RFs for cancers? The various microbial communities within a holobiont cooperate and compete with each other and with their eukaryotic host; are these interactions productively viewed as social, political, or economic, i.e., can models based on such concepts contribute predictive power that is unavailable in the languages of biochemistry, cell biology, or microbial ecology? It perhaps depends on how context-sensitive these interactions, and the RFs that support them, turn out to be.

Is it, finally, useful to think of ecosystem-scale systems or even life as a whole as "selves"? Living systems, including the living system we call "evolution", appear to minimize VFE at every scale [37]. Minimizing VFE requires detecting prediction failures. Does this require a self?

7. Conclusion: Meaning as a Multi-Scale Phenomenon

Human beings create meanings. We have argued in this paper that the creation of meanings is not unique to humans, or even to "complex" organisms, but is a ubiquitous characteristic of living systems. Descartes was wrong to view nonhumans as mindless automata: all bodies are managed by "minds" capable of specifically recognizing environmental states and, in some cases, spatially-located persistent objects, switching attention when needed, and constructing memories. The basic principles of cognition are, as suggested in [24], scale-free.

What changes, however, does this suggest for biology? The language of cognition provides abstractions, like memory and attention that generalize over wide ranges of biological phenomena. The concept of a reference frame, this time from physics, is similarly wide-ranging. The "lumping" enabled by these abstractions suggests common, general, computational mechanisms with scale-specific implementations. The often-noted structural similarities between regulatory pathways and neurofunctional networks [217], and particularly the possibility that bow-tie regulatory networks function as GNW-type broadcast architectures, suggest that such common mechanisms exist, a suggestion reinforced by the utility of active inference as an information-processing model on multiple scales.

Further development of specific abilities to manipulate existing biological RFs at the scale at which they are implemented, e.g., bioelectric manipulations of body-axis polarity in planaria [166,167] or optogenetic manipulations of particular memories in mice [49] will begin to construct a vocabulary of functional localization, analogous to the vocabulary of functional networks in the mammalian brain. The true test of these ideas, however, lies in the possibility of implementing new RFs from scratch in synthetic biological systems [216,218–221]. Paraphrasing Feynman, making it is a way of understanding it [222].

Throughout this paper, we have illustrated the use of multiple languages and techniques, at multiple scales, in organisms from microbes to humans, to understand the biology of meaning. We have also, contra [223], shown how concepts originating in the cognitive sciences apply to biological systems across the board. Disciplinary barriers, like trade barriers, benefit entrenched interests by interrupting the flow of ideas and techniques that might otherwise transform ways of thinking and experimental methods. We hope in this essay to have contributed somewhat to their dissolution.

Author Contributions: Conceptualization, C.F. and M.L.; writing—original draft preparation, C.F.; writing—review and editing, C.F. and M.L. All authors have read and agreed to the published version of the manuscript.

Funding: The work of M.L. was supported by the Barton Family Foundation and the Templeton World Charity Foundation (No. TWCF0089/AB55).

Conflicts of Interest: The authors declare no conflict of interest.

Abbreviations

The following abbreviations are used in this manuscript:

AI	Artificial Intelligence
ANN	Artificial Neural Network
GNW	Global Neuronal Workspace
GOFAI	Good Old-Fashioned AI
GRN	Gene Regulation Network
LGT	Lateral Gene Transfer
MY	Million Year
ps	picosecond (10^{-12} second)
RF	Reference Frame
VFE	Variational Free Energy

References

1. Fodor, J.A. *The Language of Thought*; Harvard University Press: Cambridge, MA, USA, 1975.
2. Hinzen, W. What is un-Cartesian linguistics? *Biolinguistics* **2014**, *8*, 226–257.
3. Evans, J.B.T. Dual processing accounts of reasoning, judgement and social cognition. *Annu. Rev. Psychol.* **2008**, *59*, 255–278. [CrossRef] [PubMed]
4. Gentner, D. Why we're so smart. In *Language and Mind: Advances in the Study of Language and Thought*; Gentner, D., Goldin-Meadow, S., Eds.; MIT Press: Cambridge, MA, USA, 2003; pp. 195–235.
5. Suddendorf, T.; Corballis, M.C. The evolution of foresight: What is mental time travel, and is it unique to humans? *Behav. Brain Sci.* **2007**, *30*, 299–351. [CrossRef]
6. Hauser, M.D.; Chomsky, N.; Fitch, W.T. The faculty of language: What is it, who has it, and how did it evolve? *Science* **2002**, *298*, 1569–1579. [CrossRef]
7. Muñoz-Dorado, J.; Marcos-Torres, F.J.; García-Bravo, E.; Moraleda-Muxnxoz, A.; Pérez, J. Myxobacteria: Moving, killing, feeding, and surviving together. *Front. Microbiol.* **2016**, *7*, 781. [CrossRef] [PubMed]
8. Calvo, P.; Kiejzer, F. Plant Cognition. In *Plant-Environment Interactions, Signaling and Communication in Plants*; Baluška, Ed.; Springer: Berlin, Germany, 2009; pp. 247–266.
9. Dreyfus, H.L. *What Computers Can't Do: A Critique of Artificial Reason*; Harper and Row: New York, NY, USA, 1972.
10. Searle, J.R. Minds, brains, and programs. *Behav. Brain Sci.* **1980**, *3*, 417–424. [CrossRef]
11. Fodor, J.A. *The Mind Doesn't Work That Way: The Scope and Limits of Computational Psychology*; MIT Press: Cambride, MA, USA, 2000.
12. Froese, T.; Taguchi, S. The problem of meaning in AI and robotics: Still with us after all these years. *Philosophies* **2019**, *4*, 14. [CrossRef]

13. Gagliano, M.; Grimonprez, M. Breaking the silence—Language and the making of meaning in plants. *Ecopsychology* **2015**, *7*, 145–152. [CrossRef]

14. Lyon, P. The cognitive cell: Bacterial behavior reconsidered. *Front. Microbiol.* **2015**, *6*, 264. [CrossRef]

15. Baluška, F.; Levin, M. On having no head: Cognition throughout biological systems. *Front. Psychol.* **2016**, *7*, 902. [CrossRef]

16. Newen, A.; De Bruin, B.; Gallagher, S. 4E cognition: Historical roots, key concepts, and central issues. In *The Oxford Handbook of 4E Cognition*; Newen, A., De Bruin, B., Gallagher, S., Eds.; Oxford University Press: Oxford, UK, 2018; pp. 3–15.

17. Kull, K.; Deacon, T.; Emmeche, C.; Hoffmeyer, J.; Stjernfelt, F. Theses on biosemiotics: Prolegomena to a theoretical biology. In *Towards a Semiotic Biology: Life Is the Action of Signs*; Emmeche, C., Kull, K., Eds.; Imperial College Press: London, UK, 2011; pp. 25–41.

18. Maturana, H.R.; Varela, F.J. *Autopoesis and Cognition: The Realization of the Living*; D. Reidel: Dordrecht, The Netherlands, 1980.

19. Varela, F.J.; Thompson, E.; Rosch, E. *The Embodied Mind*; MIT Press: Cambridge, MA, USA, 1991.

20. Pattee, H.H. Cell psychology. *Cogn. Brain Theory* **1982**, *5*, 325–341.

21. Stewart, J. Cognition = Life: Implications for higher-level cognition. *Behav. Process.* **1996**, *35*, 311–326. [CrossRef]

22. di Primio, F.; Müller, B.F.; Lengeler, J.W. Minimal cognition in unicellular organisms. In *From Animals to Animats*; Meyer, J.A., Berthoz, A., Floreano, D., Roitblat, H.L., Wilson, S.W., Eds.; International Society For Adaptive Behavior: Honolulu, HI, USA, 2000; pp. 3–12.

23. Miller, W.B.; Torday, J.S. Four domains: The fundamental unicell and post-Darwinian cognition-based evolution. *Prog. Biophys. Mol. Biol.* **2018**, *140*, 49–73. [CrossRef] [PubMed]

24. Levin, M. The computational boundary of a 'self': Developmental bioelectricity drives multicellularity and scale-free cognition. *Front. Psychol.* **2019**, *10*, 2688. [CrossRef] [PubMed]

25. Gibson, J.J. *The Ecological Approach to Visual Perception*; Houghton Mifflin: Boston, MA, USA, 1979.

26. Turvey, M.T.; Shaw, R.E.; Reed, E.S.; Mace, M.W. Ecological laws of perceiving and acting: In reply to Fodor and Pylyshyn (1981). *Cognition* **1981**, *9*, 237–304. [CrossRef]

27. Michaels, C.F.; Carello, C. *Direct Perception*; Prentice-Hall: Englewood Cliffs, NJ, USA, 1981.

28. Chemero, A. Radical embodied cognitive science. *Rev. Gen. Psychol.* **2013**, *17*, 145–150. [CrossRef]

29. Di Paolo, E.; Thompson, E. The enactive approach. In *The Routledge Handbook of Embodied Cognition*; Shapiro, L., Ed.; Routledge: London, UK, 2014; pp. 68–78.

30. Anderson, M.L. Embodied cognition: A field guide. *Artif. Intell.* **2003**, *149*, 91–130. [CrossRef]

31. Froese, T.; Ziemke, T. Enactive artificial intelligence: Investigating the systemic organization of life and mind. *Artif. Intell.* **2009**, *173*, 466–500. [CrossRef]

32. Spencer, J.P.; Perone, S.; Johnson, J.S. Dynamic field theory and embodied cognitive dynamics. In *Toward a Unified Theory of Development Connectionism and Dynamic System Theory Reconsidered*; Spencer, J.P., Ed.; Oxford University Press: Oxford, UK, 2009; pp. 86–118.

33. Clark, A. How to knit your own Markov blanket: Resisting the Second Law with metamorphic minds. In *Philosophy and Predictive Processing: 3*; Metzinger, T., Wiese, W., Eds.; MIND Group: Frankfurt am Main, Germany, 2017. [CrossRef]

34. Bbichakjian, B.H. Language: From sensory mapping to cognitive construct. *Biolinguistics* **2012**, *6*, 247–258.

35. Fields, C.; Levin, M. Multiscale memory and bioelectric error correction in the cytoplasm-cytoskeleton-membrane system. *WIRES Syst. Biol. Med.* **2018**, *10*, e1410. [CrossRef]

36. Fields, C.; Levin, M. Integrating evolutionary and developmental thinking into a scale-free biology. *BioEssays* **2020**, *42*, 1900228. [CrossRef] [PubMed]

37. Fields, C.; Levin, M. Does evolution have a target morphology? *Organisms* **2020**, *4*, 57–76.

38. Bartlett, S.D.; Rudolph, T.; Spekkens, R.W. Reference frames, superselection rules, and quantum information. *Rev. Mod. Phys.* **2007** *79*, 555–609. [CrossRef]

39. Rao, R.P.; Ballard, D.H. Predictive coding in the visual cortex: A functional interpretation of some extra-classical receptive-field effects. *Nat. Neurosci.* **1999**, *2*, 79–87. [CrossRef] [PubMed]

40. Knill, D.C.; Pouget, A. The Bayesian brain: The role of uncertainty in neural coding and computation. *Trends Neurosci.* **2004**, *27*, 712–719. [CrossRef]

41. Friston, K.J. The free-energy principle: A unified brain theory? *Nat. Rev. Neurosci.* **2010**, *11*, 127–138. [CrossRef]
42. Friston, K.J. Life as we know it. *J. R. Soc. Interface* **2013**, *10*, 20130475. [CrossRef]
43. Clark, A. Whatever next? Predictive brains, situated agents, and the future of cognitive science. *Behav. Brain Sci.* **2013**, *36*, 181–204. [CrossRef]
44. Hohwy, J. *The Predictive Mind*; Oxford University Press: Oxford, UK, 2013.
45. Spratling, M.W. Predictive coding as a model of cognition. *Cogn. Process.* **2016**, *17*, 279–305. [CrossRef]
46. Corbetta, M.; Shulman, G.L. Control of goal-directed and stimulus driven attention in the brain. *Nat. Rev. Neurosci.* **2002**, *3*, 201–215. [CrossRef]
47. Vossel, S.; Geng, J.J.; Fink, G.R. Dorsal and ventral attention systems: Distinct neural circuits but collaborative roles. *Neurosci.* **2014**, *20*, 150–159. [CrossRef] [PubMed]
48. Eichenbaum, H. Still searching for the engram. *Learn. Behav.* **2016**, *44*, 209–222. [CrossRef] [PubMed]
49. Josselyn, S.A.; Tonegawa, S. Memory engrams: Recalling the past and imagining the future. *Science* **2020**, *367*, eaaw4325. [CrossRef]
50. Nadel, L.; Hupbach, A.; Gomez, R.; Newman-Smith, K. Memory formation, consolidation and transformation. *Neurosci. Biobehav. Rev.* **2012**, *36*, 1640–1645. [CrossRef] [PubMed]
51. Schwabe, L.; Nader, K.; Pruessner, J.C. Reconsolidation of human memory: Brain mechanisms and clinical relevance. *Biol. Psychiatry* **2014**, *76*, 274–280. [CrossRef] [PubMed]
52. Craig, A.D. The sentient self. *Brain Struct. Funct.* **2010**, *214*, 563–577. [CrossRef] [PubMed]
53. Northoff, G. Brain and self—A neurophilosophical account. *Child Adolesc. Psychiatry Ment. Health* **2013**, *7*, 28. [CrossRef]
54. Seth, A.K. Interoceptive inference, emotion, and the embodied self. *Trends Cogn. Sci.* **2013**, *17*, 565–573. [CrossRef]
55. Seth, A.K.; Tsakiris, M. Being a beast machine: The somatic basis of selfhood. *Trends Cogn. Sci.* **2018**, *22*, 969–981. [CrossRef]
56. Bateson, G. *Steps to an Ecology of Mind: Collected Essays in Anthropology, Psychiatry, Evolution, and Epistemology*; Jason Aronson: Northvale, NJ, USA, 1972.
57. Roederer, J. *Information and Its Role in Nature*; Springer: Berlin, Germany, 2005.
58. Darwin, C. *On the Origin of Species by Means of Natural Selection, or the Preservation of Favoured Races in the Struggle for Life*; Murray: London, UK, 1859. Available online: http://darwin-online.org.uk/Variorum/1859 (accessed on 9 November 2020).
59. Schrodinger, E. *What is Life?* Cambridge University Press: Cambridge, UK, 1944.
60. Lovelock, J.E.; Margulis, L. Atmospheric homeostasis by and for the biosphere: The Gaia hypothesis. *Tellus* **1974**, *26*, 2–10. [CrossRef]
61. Kauffman, S.A. *The Origins of Order: Self Organization and Selection in Evolution*; Oxford University Press: Oxford, UK, 1993.
62. Bartlett, S.; Wong, M.L. Defining Lyfe in the Universe: From three privileged functions to four pillars. *Life* **2020**, *10*, 42. [CrossRef]
63. Hermida, M. Life on Earth is an individual. *Theory Biosci.* **2016**, *135*, 37–44. [CrossRef] [PubMed]
64. Mariscal, C.; Doolittle, W.F. Life and life only: A radical alternative to life definitionism. *Synthese* **2020**, *197*, 2975–2989. [CrossRef]
65. Fields, C.; Marcianò, A. Markov blankets are general physical interaction surfaces. *Phys. Life Rev.* **2020**, *33*, 109–111. [CrossRef] [PubMed]
66. Fields, C.; Glazebrook, J.F. Representing measurement as a thermodynamic symmetry breaking. *Symmetry* **2019**, *12*, 810. [CrossRef]
67. Fields, C.; Marcianò, A. Holographic screens are classical information channels. *Quantum Rep.* **2020**, *2*, 326–336. [CrossRef]
68. Landauer, R. Irreversibility and heat generation in the computing process. *IBM J. Res. Devel.* **1961**, *5*, 183–195. [CrossRef]
69. Parrondo, J.M.R.; Horowitz, J.M.; Sagawa, T. Thermodynamics of information. *Nat. Phys.* **2015**, *11*, 131–139. [CrossRef]
70. Moore, E.F. Gedankenexperiments on sequential machines. In *Autonoma Studies*; Shannon, C.W., McCarthy, J., Eds.; Princeton University Press: Princeton, NJ, USA, 1956; pp. 129–155.

71. Fields, C. Some consequences of the thermodynamic cost of system identification. *Entropy* **2018**, *20* 797. [CrossRef]

72. Fields, C. Building the observer into the system: Toward a realistic description of human interaction with the world. *Systems* **2016**, *4*, 32. [CrossRef]

73. Fields, C. Sciences of observation. *Philosophies* **2018**, *3*, 29. [CrossRef]

74. Friston, K.; Levin, M.; Sengupta, B.; Pezzulo, G. Knowing one's place: A free-energy approach to pattern regulation. *J. R. Soc. Interface* **2015**, *12*, 20141383. [CrossRef] [PubMed]

75. Kuchling, F.; Friston, K.; Georgiev, G.; Levin, M. Morphogenesis as Bayesian inference: A variational approach to pattern formation and control in complex biological systems. *Phys. Life Rev.* **2020**, *33*, 88–108. [CrossRef] [PubMed]

76. Hume, D. *An Enquiry Concerning Human Understanding*; A. Millar: London, UK, 1748. Available online: http://www.gutenberg.org/ebooks/9662 (accessed on 9 November 2020).

77. Pattee, H.H. The physics of symbols: Bridging the epistemic cut. *Biosystems* **2001**, *60*, 5–21. [CrossRef]

78. Micali, G.; Endres, R.G. Bacterial chemotaxis: Information processing, thermodynamics, and behavior. *Curr. Opin. Microbiol.* **2016**, *30*, 8–15. [CrossRef]

79. Tweedy, L.; Thomason, P.A.; Paschke, P.I.; Martin, K.; Machesky, L.M.; Zagnoni, M.; Insall, R.H. Seeing around corners: Cells solve mazes and respond at a distance using attractant breakdown. *Science* **2020**, *369*, eaay9792. [CrossRef]

80. Baron, S.; Eisenbach, M. CheY acetylation is required for ordinary adaptation time in *Escherichia coli* chemotaxis. *FEBS Lett.* **2017**, *591*, 1958–1965. [CrossRef]

81. Fields, C.; Marcianò, A. Sharing nonfungible information requires shared nonfungible information. *Quant. Rep.* **2019**, *1*, 252–259. [CrossRef]

82. Herculano-Houzel, S. Scaling of brain metabolism with a fixed energy budget per neuron: Implications for neuronal activity, plasticity and evolution. *PLoS ONE* **2011**, *6*, e17514. [CrossRef]

83. Robbins, R.J.; Krishtalka, L.; Wooley, J.C. Advances in biodiversity: Metagenomics and the unveiling of biological dark matter. *Stand. Genom. Sci.* **2016**, *11*, 69. [CrossRef]

84. Cabezón, E.; Ripoll-Rozada, J.; Peña, A.; de la Cruz, F.; Arechaga, I. Towards an integrated model of bacterial conjugation. *FEMS Microbiol. Rev.* **2015**, *39*, 81–95. [CrossRef] [PubMed]

85. Chen, A.H.; Lubkowicz, D.; Yeong, V.; Chang, R.L.; Silver, P.A. Transplantability of a circadian clock to a noncircadian organism. *Sci. Adv.* **2015**, *1*, e1500358. [CrossRef] [PubMed]

86. Barkow, J.; Cosmides, L.; Tooby, J. *The Adapted Mind: Evolutionary Psychology and the Generation of Culture*; Oxford University Press: Oxford, UK, 1992.

87. Buss, D.M. (Ed.) *The Handbook of Evolutionary Psychology*; John Wiley: Hoboken, NJ, USA, 2005.

88. Cook, N.D.; Carvalho, G.B.; Damasio, A. From membrane excitability to metazoan psychology. *Trends Neurosci.* **2014**, *37*, 698–705. [CrossRef] [PubMed]

89. Capra, E.J.; Laub, M.T. Evolution of two-component signal transduction systems. *Annu. Rev. Microbiol.* **2012**, *66*, 325–347. [CrossRef] [PubMed]

90. Loh, K.M.; van Amerongen, R.; Nusse, R. Generating cellular diversity and spatial form: Wnt signaling and the evolution of multicellular animals. *Dev. Cell* **2016**, *38*, 643–655. [CrossRef] [PubMed]

91. Li, M.; Liu, J.; Zhang, C. Evolutionary history of the vertebrate mitogen activated protein kinases family. *PLoS ONE* **2011**, *6*, e26999. [CrossRef]

92. Fischer, A.H.L.; Smith, J. Evo–Devo in the era of gene regulatory networks. *Integr. Comp. Biol.* **2012**, *52*, 842–849. [CrossRef]

93. Nickel, M. Evolutionary emergence of synaptic nervous systems: What can we learn from the non-synaptic, nerveless Porifera? *Invertebr. Biol.* **2010**, *129*, 1–16. [CrossRef]

94. Roshchina, V.V. New trends and perspectives in the evolution of neurotransmitters in microbial, plant, and animal cells. In *Microbial Endocrinology: Interkingdom Signaling in Infectious Disease and Health*; Lyte, M., Ed.; Springer: Cham, Switzerland, 2016; pp. 25–77.

95. Csaba, G. The hormonal system of the unicellular Tetrahymena: A review with evolutionary aspects. *Acta Microbiol. Immunol. Hungarica* **2012**, *59*, 131–156. [CrossRef]

96. Campbell, R.K.; Satoh, N.; Degnan, B.M. Piecing together evolution of the vertebrate endocrine system. *Trends Genet.* **2004**, *20*, 359–366. [CrossRef]

97. Levin, M. Morphogenetic fields in embryogenesis, regeneration, and cancer: Non-local control of complex patterning. *Biosystems* **2012**, *109*, 243–261. [CrossRef] [PubMed]

98. Levin, M. Molecular bioelectricity: How endogenous voltage potentials control cell behavior and instruct pattern regulation in vivo. *Mol. Biol. Cell.* **2014**, *25*, 3835–3850 [CrossRef] [PubMed]

99. Levin, M.; Martyniuk, C.J. The bioelectric code: An ancient computational medium for dynamic control of growth and form. *BioSystems* **2018**, *164*, 76–93. [CrossRef]

100. Arendt, D.; Tosches, M.A.; Marlow, H. From nerve net to nerve ring, nerve cord and brain—Evolution of the nervous system. *Nat. Rev. Neurosci.* **2016**, *17*, 61–72. [CrossRef] [PubMed]

101. Varoqueaux, F.; Fasshauer D. Getting nervous: An evolutionary overhaul for communication. *Annu. Rev. Genet.* **2017**, *51*, 455–476. [CrossRef]

102. Fields, C.; Bischof, J.; Levin, M. Morphological coordination: A common ancestral function unifying neural and non-neural signaling. *Physiology* **2020**, *35*, 16–30. [CrossRef]

103. Guerrero, R.; Margulis, L.; Berlanga, M. Symbiogenesis: The holobiont as a unit of evolution. *Int. Microbiol.* **2013**, *16*, 133–143.

104. Gilbert, S.F. Symbiosis as the way of eukaryotic life: The dependent co-origination of the body. *J. Biosci.* **2014**, *39*, 201–209. [CrossRef]

105. Thiery, S.; Kaimer, C. The predation strategy of *Myxococcus Xanthus*. *Front. Microbiol.* **2020**, *11*, 2. [CrossRef]

106. Turner, J.S. Extended phenotypes and extended organisms. *Biol. Philos.* **2004**, *19*, 327–352. [CrossRef]

107. Schultz, T.R.; Brady, S.G. Major evolutionary transitions in ant agriculture. *Proc. Natl. Acad. Sci. USA* **2008**, *105*, 5435–5440. [CrossRef]

108. Rakison, D.H.; Yermolayeva, Y. Infant categorization. *WIRES Cogn. Sci.* **2010**, *1*, 894–905. [CrossRef]

109. Baillargeon, R.; Stavans, M.; Wu, D.; Gertner, Y.; Setoh, P.; Kittredge, A.K.; Bernard, A. Object individuation and physical reasoning in infancy: An integrative account. *Lang. Learn. Dev.* **2012**, *8*, 4–46. [CrossRef] [PubMed]

110. Yan, D.; Lin, X. Shaping morphogen gradients by proteoglycans. *Cold Spring Harb. Perspect. Biol.* **2009**, *1*, a002493. [CrossRef] [PubMed]

111. Clause, K.C.; Barker, T.H. Extracellular matrix signaling in morphogenesis and repair. *Curr. Opin. Biotechnol.* **2013**, *24*, 830–833. [CrossRef] [PubMed]

112. Gogna, R.; Shee, K.; Moreno, E. Cell competition during growth and regeneration. *Annu. Rev. Genet.* **2015**, *49*, 697–718. [CrossRef] [PubMed]

113. Madan, E.; Gogna, R.; Moreno, E. Cell competition in development: Information from flies and vertebrates. *Curr. Opin. Cell Biol.* **2018**, *55*, 150–157. [CrossRef] [PubMed]

114. Martin, A. The representation of object concepts in the brain. *Annu. Rev. Psychol.* **2007**, *58*, 25–45. [CrossRef]

115. Zimmer, H.D.; Ecker, U.K.H. Remembering perceptual features unequally bound in object and episodic tokens: Neural mechanisms and their electrophysiological correlates. *Neurosci. Biobehav. Rev.* **2010**, *34*, 1066–1079. [CrossRef]

116. Keifer, M.; Pulvermüller, F. Conceptual representations in mind and brain: Theoretical developments, current evidence and future directions. *Cortex* **2012**, *7*, 805–825. [CrossRef]

117. Clarke, A.; Tyler, L.K. Understanding what we see: How we derive meaning from vision. *Trends Cogn. Sci.* **2015**, *19*, 677–687. [CrossRef] [PubMed]

118. Fields, C. Visual re-identification of individual objects: A core problem for organisms and AI. *Cogn. Process.* **2016**, *17*, 1–13. [CrossRef] [PubMed]

119. Yau, J.M.; Kim, S.S.; Thakur, P.H.; Bensmaia, S.J. Feeling form: The neural basis of haptic shape perception. *J. Neurophysiol.* **2016**, *115*, 631–642. [CrossRef] [PubMed]

120. Fields, C. Object Permanence. In *Encyclopedia of Evolutionary Psychological Science*; Shackelford, T.K., Weekes-Shackelford, V.A., Eds.; Springer: New York, NY, USA, 2017; Chapter 2373.

121. Eichenbaum, H.; Yonelinas, A.R.; Ranganath, C. The medial temporal lobe and recognition memory. *Annu. Rev. Neurosci.* **2007**, *30*, 123–152. [CrossRef] [PubMed]

122. Fields, C. The very same thing: Extending the object token concept to incorporate causal constraints on individual identity. *Adv. Cogn. Psychol.* **2012**, *8*, 234–247. [CrossRef] [PubMed]

123. Pitts, W.; McCulloch, W.S. How we know universals: The perception of auditory and visual forms. *Bull. Math. Biophys.* **1947**, *9*, 127–147. [CrossRef]

124. Sasaki, Y.; Vuffel, W.; Knutsen, T.; Tyler, C.; Tootell, R. Symmetry activates extrastriate visual cortex in human and nonhuman primates. *Proc. Natl. Acad. Sci. USA* **2005**, *102*, 3159–3163. [CrossRef]

125. Fields, C. How humans solve the frame problem. *J. Expt. Theor. Artif. Intell.* **2013**, *25*, 441–456. [CrossRef]

126. Dietrich, E.; Fields, C. Equivalence of the Frame and Halting problems. *Algorithms* **2020**, *13*, 175. [CrossRef]

127. Wigner, E.P. Remarks on the mind-body question. In *The Scientist Speculates*; Good, I.J., Ed.; Heinemann: London, UK, 1961; pp. 284–302.

128. Schwartz, J.M.; Stapp, H.P.; Beauregard, M. Quantum physics in neuroscience and psychology: A neurophysical model of mind-brain interaction. *Philos. Trans. R. Soc. B* **2005**, *360*, 1309–1327. [CrossRef]

129. Hameroff, S.; Penrose, R. Consciousness in the universe: A review of the 'OrchOR' theory. *Phys. Life Rev.* **2014**, *11*, 39–78. [CrossRef]

130. Fields, C. If physics is an information science, what is an observer? *Information* **2012**, *3* 92–123. [CrossRef]

131. Niehrs, C. On growth and form: A Cartesian coordinate system of Wnt and BMP signaling specifies bilaterian body axes. *Development* **2010**, *137*, 845–857. [CrossRef] [PubMed]

132. Chichili, G.R.; Rodgers, W. Cytoskeleton-membrane interactions in membrane raft structure. *Cell. Mol. Life Sci.* **2009**, *66*, 2319–2328. [CrossRef] [PubMed]

133. Gidon, A.; Zolnik, T.A.; Fidzinski, P.; Bolduan, F.; Papoutsi, A.; Panaylota, P.; Holtkamp, M.; Vida, I.; Larkum, M.E. Dendritic action potentials and computation in human layer 2/3 cortical neurons. *Science* **2020**, *367*, 83–87. [CrossRef] [PubMed]

134. Penfield, W.; Boldrey, E. Somatic motor and sensory representation in the cerebral cortex of man as studied by electrical stimulation. *Brain* **1937**, *60*, 389–443. [CrossRef]

135. Gardner, J.L.; Merriam, E.P.; Movshon, J.A.; Heeger, D.J. Maps of visual space in human occipital cortex are retinotopic, not spatiotopic. *J. Neurosci.* **2008**, *28*, 3988–3999. [CrossRef]

136. Moser, E.I.; Kropff, E.; Moser, M.-B. Place cells, grid cells, and the brain's spatial representation system. *Annu. Rev. Neurosci.* **2008**, *31*, 31–89. [CrossRef]

137. Abrams, N.E.; Primack, J.R. Cosmology and 21st century culture. *Science* **2001**, *293*, 1769–1770. [CrossRef]

138. Johnson, C.H. Precise circadian clocks in prokaryotic cyanobacteria. *Curr. Issues Mol. Biol.* **2004**, *6*, 103–110.

139. Doherty, C.J.; Kay, S.A. Circadian control of global gene expression patterns. *Annu. Rev. Genet.* **2010**, *44*, 419–444. [CrossRef] [PubMed]

140. Chakrabarti, S.; Michor, F. Circadian clock effects on cellular proliferation: Insights from theory and experiments. *Curr. Opin. Cell Biol.* **2020**, *67*, 17–26. [CrossRef] [PubMed]

141. Kuchen, E.E.; Becker, N.B.; Claudino, N.; Höfer, T. Hidden long-range memories of growth and cycle speed correlate cell cycles in lineage trees. *eLife* **2020**, *9*, e51002. [CrossRef] [PubMed]

142. Beck, C.W.; Izpisúa Belmonte, J.C.I.; Christen, B. Beyond early development: *Xenopus* as an emerging model for the study of regenerative mechanisms. *Dev. Dyn.* **2009**, *238*, 1226–1248. [CrossRef] [PubMed]

143. Friston, K.; Rigoli, F.; Ognibene, D.; Mathys, C.; FitzGerald, T.; Pezzulo, G. Active inference and epistemic value. *Cogn. Neurosci.* **2015**, *6*, 187–214. [CrossRef] [PubMed]

144. Godfrey-Smith, P. *Other Minds: The Octopus, the Sea, and the Deep Origins of Consciousness*; Farrar, Straus and Giroux: New York, NY, USA, 2016.

145. Wang, D.; Jin, S.; Zou, X. Crosstalk between pathways enhances the controllability of signalling networks. *IET Syst. Biol.* **2016**, *10*, 2–9. [CrossRef] [PubMed]

146. Brodskiy, P.A.; Zartmann, J.J. Calcium as a signal integrator in developing epithelial tissues. *Phys. Biol.* **2018**, *15*, 051001. [CrossRef]

147. Niss, K.; Gomez-Casado, C.; Hjaltelin, J.X.; Joeris, T.; Agace, W.W.; Belling, K.G.; Brunak, S. Complete topological mapping of a cellular protein interactome reveals bow-tie motifs as ubiquitous connectors of protein complexes. *Cell Rep.* **2020**, *31*, 107763. [CrossRef]

148. Dehaene, S.; Naccache, L. Towards a cognitive neuroscience of consciousness: Basic evidence and a workspace framework. *Cognition* **2001**, *79*, 1–37. [CrossRef]

149. Baars, B.J.; Franklin, S. How conscious experience and working memory interact. *Trends Cogn. Sci.* **2003**, *7*, 166–172. [CrossRef]

150. Baars, B.J.; Franklin, S.; Ramsoy, T.Z. Global workspace dynamics: Cortical "binding and propagation" enables conscious contents. *Front. Psychol.* **2013**, *4*, 200. [CrossRef] [PubMed]

151. Dehaene, S.; Charles, L.; King, J.-R.; Marti, S. Toward a computational theory of conscious processing. *Curr. Opin. Neurobiol.* **2014**, *25*, 76–84. [CrossRef] [PubMed]

152. Mashour, G.A.; Roelfsema, P.; Changeux, J.-P.; Dehaene, S. Conscious processing and the Global Neuronal Workspace hypothesis. *Neuron* **2020**, *105*, 776–798. [CrossRef] [PubMed]
153. Jacob, F.; Monod, J. Genetic regulatory mechanisms in the synthesis of proteins. *J. Mol. Biol.* **1961**, *3*, 318–356. [CrossRef]
154. Dzhafarov, E.N.; Kujala, J.V.; Cervantes, V.H. Contextuality-by-default: A brief overview of concepts and terminology. In *Lecture Notes in Computer Science 9525*; Atmanspacher, H., Filik, T., Pothos, E., Eds.; Springer: Berlin, Germeny, 2016; pp. 12–23.
155. Dzhafarov, E.N.; Cervantes, V.H.; Kujala, J.V. Contextuality in canonical systems of random variables. *Philos. Trans. R. Soc. A* **2017**, *375*, 20160389. [CrossRef] [PubMed]
156. Dzharfarov, E.N.; Kon, M. On universality of classical probability with contextually labeled random variables. *J. Math. Psych.* **2018**, *85*, 17–24. [CrossRef]
157. Mermin, N.D. Hidden variables and the two theorems of John Bell. *Rev. Mod. Phys.* **1993**, *65*, 803–815. [CrossRef]
158. Fields, C.; Glazebrook, J.F. Information flow in context-dependent hierarchical Bayesian inference. *J. Expt. Theor. Artif. Intell.* **2020**. [CrossRef]
159. Pietsch, T.W.; Grobecker, D.B. The compleat angler: Aggressive mimicry in an Antennariid anglerfish. *Science* **1978**, *201*, 369–370. [CrossRef]
160. Sowa, J.F. (Ed.) *Principles of Semantic Networks: Explorations in the Representation of Knowledge*; Morgan Kauffman: San Mateo, CA, USA, 2014.
161. Uddin, L.Q. Salience processing and insular cortical function and dysfunction. *Nat. Rev. Neurosci.* **2015**, *16*, 55–61. [CrossRef]
162. Schaefer, H.M.; Ruxton, G.D. Signal diversity, sexual selection and speciation. *Annu. Rev. Ecol. Evol. Syst.* **2015**, *46*, 573–592. [CrossRef]
163. Müller, G.B. Evo-devo: Extending the evolutionary synthesis. *Nat. Rev. Genet.* **2007**, *8*, 943–949. [CrossRef] [PubMed]
164. Carroll, S.B. Evo-devo and an expanding evolutionary synthesis: A genetic theory of morphological evolution. *Cell* **2008**, *134*, 25–36. [CrossRef] [PubMed]
165. Sardet, C.; Paix, A.; Prodon, F.; Dru, P.; Chenevert, J. From oocyte to 16-cell stage: Cytoplasmic and cortical reorganizations that pattern the ascidian embryo. *Dev. Dyn.* **2007**, *36*, 1716–1731. [CrossRef] [PubMed]
166. Durant, F.; Morokuma, J.; Fields, C.; Williams, K.; Adams, D.S.; Levin, M. Long-term, stochastic editing of regenerative anatomy via targeting endogenous bioelectric gradients. *Biophys. J.* **2017**, *112*, 2231–2243. [CrossRef] [PubMed]
167. Durant, F.; Bischof, J.; Fields, C.; Morokuma, J.; LaPalme, J.; Hoi, A.; Levin, M. The role of early bioelectric signals in the regeneration of planarian anterior-posterior polarity. *Biophys. J.* **2019**, *116*, 948–961. [CrossRef]
168. Pietak, A.; Bischof, J.; LaPalme, J.; Morokuma, J.; Levin, M. Neural control of body-plan axis in regenerating planaria. *PLoS Comput. Biol.* **2019**, *15*, e1006904. [CrossRef]
169. Armus, H.L.; Montgomery, A.R.; Jellison, J.L. Discrimination learning in paramecia (*P. caudatum*). *Psychol. Rec.* **2006**, *56*, 489–498. [CrossRef]
170. Shirakawa, T.; Gunji, Y.-P.; Miyake, Y. An associative learning experiment using the plasmodium of *Physarum polycephalum*. *Nano Commun. Net.* **2011**, *2*, 99–105. [CrossRef]
171. Karpiński, S.; Szechyxnxska-Hebda, M. Secret life of plants: From memory to intelligence. *Plant Signal Behav.* **2010**, *5*, 1391–1394. [CrossRef]
172. Abramson, C.I.; Chicas-Mosier, A.M. Learning in plants: Lessons from *Mimosa pudica*. *Front. Psychol.* **2016**, *7*, 417. [CrossRef] [PubMed]
173. Gagliano, M.; Vyazovskiy, V.V.; Borbély, A.A.; Grimonprez, M.; Depczynski, M. Learning by association in plants. *Sci. Rep.* **2016**, *6*, 38427. [CrossRef] [PubMed]
174. Lee, S.W.; O'Doherty, J.P.; Shimojo, S. Neural computations mediating one-shot learning in the human brain. *PLoS Biol.* **2015**, *13*, e1002137. [CrossRef] [PubMed]
175. Major, G.; Larkum, M.E.; Schiller, J. Active properties of neocortical pyramidal neuron dendrites. *Annu. Rev. Neurosci.* **2013**, *36*, 1–24. [CrossRef]
176. Eyal, G.; Verhoog, M.B.; Test-Silva, G.; Deitcher, Y.; Benavides-Piccione, R.; DeFelipe, J.; de Kock, C.P.J.; Mansvelder, H.D.; Segev, I. Human cortical pyramidal neurons: From spines to spikes via models. *Front. Cell. Neurosci.* **2018**, *12*, 181. [CrossRef]

177. Day, J.J.; Sweatt, J.D. Cognitive neuroepigenetics: A role for epigenetic mechanisms in learning and memory. *Neurobiol. Learn. Mem.* **2011**, *96*, 2–12. [CrossRef]

178. Marshall, P.; Bredy, T.W. Cognitive neuroepigenetics: The next evolution in our understanding of the molecular mechanisms underlying learning and memory? *NPJ Sci. Learn.* **2016**, 16014. [CrossRef]

179. Schuman, C.D.; Potok, T.E.; Patton, R.M.; Birdwell, D.; Dean, M.E.; Rose, G.S.; Plank, J.S. A survey of neuromorphic computing and neural networks in hardware. *arXiv* **2017**, arXiv:1705.06963v1.

180. Foster, P.L. Adaptive mutation: Implications for evolution. *BioEssays* **2000**, *22*, 1067–1074. [CrossRef]

181. Mitchell, A.; Romano, G.H.; Groisman, B.; Yona, A.; EDekel, E.; Kupiec, M.; Dahan, O.; Pilpel, Y. Adaptive prediction of environmental changes by microorganisms. *Nature* **2009**, *460*, 220–224. [CrossRef]

182. Corballis, M.C. The evolution of language. *Proc. N. Y. Acad. Sci.* **2009**, *1156*, 19–43. [CrossRef] [PubMed]

183. Bro-Jorgensen, J. Dynamics of multiple signalling systems: Animal communication in a world in flux. *Trends Ecol. Evol.* **2010**, *25*, 292–300. [CrossRef] [PubMed]

184. Hebets, E.A.; Barron, A.B.; Balakrishnan, C.N.; Hauber, M.E.; Mason, P.H.; Hoke, K.L. A systems approach to animal communication. *Proc. R. Soc. B* **2016**, *283*, 20152889. [CrossRef] [PubMed]

185. Iwaniuk, A.N.; Lefebvre, L.; Wylie, D.R. The comparative approach and brain-behaviour relationships: A tool for understanding tool use. *Can. J. Exp. Psychol.* **2009**, *63*, 150–159. [CrossRef] [PubMed]

186. Lefebvre, L. Brains, innovations, tools and cultural transmission in birds, non-human primates, and fossil hominins. *Front. Hum. Neurosci.* **2013**, *7*, 245. [CrossRef]

187. McGrew, W.C. Is primate tool use special? Chimpanzee and New Caledonian crow compared. *Philos. Trans. R. Soc. B* **2013**, *368* 20120422. [CrossRef]

188. Navarrete, A.F.; Reader, S.M.; Street, S.E.; Whalen, A.; Lal, K.N. The coevolution of innovation and technical intelligence in primates. *Philos. Trans. R. Soc. B* **2016**, *371*, 20150186. [CrossRef]

189. Fisher, S.E.; Scharff, C. FOXP2 as a molecular window into speech and language. *Trends Genet.* **2009**, *25*, 166–177. [CrossRef]

190. Adolphs, R. The social brain: Neural basis of social knowledge. *Annu. Rev. Psychol.* **2009**, *60*, 693–716. [CrossRef]

191. Blackiston, D.J.; Shomrat, T.; Levin, M. The stability of memories during brain remodeling: A perspective. *Commun. Integr. Biol.* **2015**, *8*, e1073424. [CrossRef]

192. Henriques, G. The Tree of Knowledge system and the theoretical unification of psychology. *Rev. Gen. Psychol.* **2003**, *7*, 150–182. [CrossRef]

193. Mercier, H.; Sperber, D. Why do humans reason? Arguments for an argumentative theory. *Behav. Brain Sci.* **2011**, *34*, 57–111. [CrossRef] [PubMed]

194. Trivers, R.L. *The Folly of Fools: The Logic of Deceit and Self-Deception in Human Life*; Basic Books: New York, NY, USA, 2011.

195. Cushman, F. Rationalization is rational. *Behav. Brain Sci.* **2020**, *43*, e28. [CrossRef] [PubMed]

196. Bargh, J.A.; Ferguson, M.J. Beyond behaviorism: On the automaticity of higher mental processes. *Psychol. Bull.* **2000**, *126*, 925–945. [CrossRef]

197. Bargh, J.A.; Schwader, K.L.; Hailey, S.E.; Dyer, R.L.; Boothby, E.J. Automaticity in social-cognitive processes. *Trends Cogn. Sci.* **2012**, *16*, 593–605. [CrossRef]

198. Csikszentmihályi, M. *Flow: The Psychology of Optimal Experience*; Harper and Row: New York, NY, USA, 1990.

199. Melnikoff, D.E.; Bargh, J.A. The mythical number two. *Trends Cogn. Sci.* **2018**, *22*, 280–293. [CrossRef]

200. Chater, N. *The Mind is Flat. The Remarkable Shallowness of the Improvising Brain*; Allen Lane: London, UK, 2018.

201. Hoffman, D.D.; Singh, M.; Prakash, C. The interface theory of perception. *Psychon. Bull. Rev.* **2015**, *22*, 1480–1506. [CrossRef]

202. Fields, C.; Glazebrook, J.F. Do Process-1 simulations generate the epistemic feelings that drive Process-2 decision-making? *Cogn. Process.* **2020**. [CrossRef]

203. Gallistel, C.R. The neurobiological bases for the computational theory of mind. In *On Concepts, Modules, and Language*; de Almeida, R.G., Gleitman, L., Eds.; Oxford University Press: New York, NY, USA, 2017; pp. 275–296.

204. Chomsky, N. Review of B. F. Skinner, Verbal Behavior. *Language* **1959**, *35*, 26–58. [CrossRef]

205. Martins, M.J.D.; Muršič, Z.; Oh, J.; Fitch, W.T. Representing visual recursion does not require verbal or motor resources. *Cogn. Psych.* **2015**, *77*, 20–41. [CrossRef]

206. Vicari, G.; Adenzato, M. Is recursion language-specific? Evidence of recursive mechanisms in the structure of intentional action. *Conscious. Cogn.* **2014**, *26*, 169–188. [CrossRef] [PubMed]
207. Martins, M.J.D.; Blanco, R.; Sammler, D.; Villringer, A. Recursion in action: An fMRI study on the generation of new hierarchical levels in motor sequences. *Hum. Barin Mapp.* **2019**, *40*, 2623–2638. [CrossRef] [PubMed]
208. Christiansen, M.H.; Chater, N. The language faculty that wasn't: A usage-based account of natural language recursion. *Front. Psych.* **2015**, *6*, 1182. [CrossRef] [PubMed]
209. Bubic, A.; von Cramon, D.Y.; Schubotz, R. Prediction, cognition and the brain. *Front. Hum. Neurosci.* **2010**, *4*, 25. [CrossRef] [PubMed]
210. Shipp, S.; Adams, R.A.; Friston, K.J. Reflections on agranular architecture: Predictive coding in the motor cortex. *Trends Neurosci.* **2013**, *36*, 706–716. [CrossRef]
211. Fields, C. Metaphorical motion in mathematical reasoning: Further evidence for pre-motor implementation of structure mapping in abstract domains. *Cogn. Process.* **2013**, *14*, 217–229. [CrossRef]
212. Amalric, M.; Dehaene, S. Origins of the brain networks for advanced mathematics in expert mathematicians. *Proc. Natl. Acad. Sci. USA* **2016**, *113*, 4909–4917. [CrossRef]
213. Fingelkurts, A.A.; Fingelkurts, A.A.; Kallio-Tamminen, T. Selfhood triumvirate: From phenomenology to brain activity and back again. *Conscious. Cogn.* **2020**, *86*, 103031.
214. Metzinger, T. *Being No One: The Self-Model Theory of Subjectivity*; MIT Press: Cambridge, MA, USA, 2003.
215. Graziano, M.S.A.; Webb, T.W. A mechanistic theory of consciousness. *Int. J. Mach. Conscious.* **2014**, *6*, 163–176. [CrossRef]
216. Levin, M. Life, death, and self: Fundamental questions of primitive cognition viewed through the lens of body plasticity and synthetic organisms. *Biochem. Biophys. Res. Commun.* **2020**. [CrossRef]
217. Barabási, A.-L. Scale-free networks: A decade and beyond. *Science* **2009**, *325*, 412–413. [CrossRef] [PubMed]
218. Solé, R.V.; Macia, J. Expanding the landscape of biological computation with synthetic multicellular consortia. *Nat. Comput.* **2013**, *12*, 485–497. [CrossRef]
219. Kamm, R.D.; Bashir, R. Creating living cellular machines. *Ann. Biomed. Eng.* **2014**, *42*, 445–459. [CrossRef] [PubMed]
220. Kriegman, S.; Cheney, N.; Bongard, J. How morphological development can guide evolution. *Sci. Rep.* **2018**, *8*, 13934. [CrossRef] [PubMed]
221. Kriegman, S.; Blackiston, D.; Levin, M.; Bongard, J. A scalable pipeline for designing reconfigurable organisms. *Proc. Natl. Acad. Sci. USA* **2020**, *117*, 1853–1859. [CrossRef]
222. Way, M. What I cannot create, I do not understand. *J. Cell Sci.* **2017**, *130*, 2941–2942. [CrossRef]
223. Fodor, J.A. Why paramecia don't have mental representations. *Midwest Stud. Philos.* **1986**, *10*, 3–23. [CrossRef]

MDPI
St. Alban-Anlage 66
4052 Basel
Switzerland
Tel. +41 61 683 77 34
Fax +41 61 302 89 18
www.mdpi.com

Philosophies Editorial Office
E-mail: philosophies@mdpi.com
www.mdpi.com/journal/philosophies